Peak Flow Measurement

Peak Flow Measurement

An illustrated guide

J.G. Ayres
Birmingham Heartlands Hospital
Birmingham
UK

and

P.J. Turpin
Yardley Green Medical Centre
Birmingham
UK

CHAPMAN & HALL MEDICAL

London · Weinheim · New York · Tokyo · Melbourne · Madras

Published by Chapman & Hall, 2–6 Boundary Row, London SE1 8HN, UK

Chapman & Hall, 2–6 Boundary Row, London SE1 8HN, UK

Blackie Academic & Professional, Wester Cleddens Road, Bishopbriggs, Glasgow G64 2NZ, UK

Chapman & Hall GmbH, Pappelallee 3, 69469 Weinheim, Germany

Chapman & Hall USA, 115 Fifth Avenue, New York, NY 10003, USA

Chapman & Hall Japan, ITP-Japan, Kyowa Building, 3F, 2-2-1 Hirakawacho, Chiyoda-ku, Tokyo 102, Japan

Chapman & Hall Australia, 102 Dodds Street, South Melbourne, Victoria 3205, Australia

Chapman & Hall India, R. Seshadri, 32 Second Main Road, CIT East, Madras 600 035, India

First edition 1997

Typeset in 12/14 pt Garamond 3 by Best-set Typesetter Ltd., Hong Kong

Printed in Italy at the Vincenzo Bona s.r.l. - Torino

ISBN 0 412 73620 9

A catalogue record for this book is available from the British Library

Library of Congress Catalog Card Number: 96-83632

∞ Printed on acid-free text paper, manufactured in accordance with ANSI/NISO Z39.48-1992 (Permanence of Paper).

Contents

Acknowledgements

We owe a great debt of thanks to Sue and Debbie, not only for their patience but also for their unstinting support.

Mavis Jervis has produced endless repeat typescripts at speed and with consummate accuracy. To her we are particularly grateful. Judith Williams helped greatly in attending to our paediatric shortcomings and Bob Torrance to our linguistic quirks. To both we are most grateful.

We wish to acknowledge our thanks to the editors of various journals and the authors of specific papers for permission to publish the following figures:

British Journal of Diseases of the Chest	Fig. A.11
British Medical Journal	Figs 1.10, 2.3, A.10, A.12
European Journal of Respiratory Diseases	Figs 4.4, 4.5, 4.12, 4.13
Thorax	Figs 3.26, 4.6, 4.8, 4.9, A.13
Clinical & Experimental Allergy	Fig. 5.3
Schweitze Medzinische Wochenschrift	Figs A.1, A.2
Respiratory Medicine	Figs A.9, A.14

Our thanks also go to Professor C.A.C. Pickering for permission to reproduce Fig. 4.10.

Particular thanks are due to Annalisa Page of Chapman & Hall who took us on and to Joanna Koster who guided us through the final stages of production.

J.A.

P.T.

Introduction

> 'I was born with inflammation of the lungs, and of everything else, I believe, that was capable of inflammation,' returned Mr Bounderby. 'For years, ma'am, I was one of the most miserable little wretches ever seen.'
>
> Charles Dickens, *Hard Times*, Book 1, ch. 4

Asthma is a common condition. It affects at least 12% of school children and 5% of adults in the UK. This amounts to around 140 patients in a General Practitioner's average list of 2000, not all of whom can be managed by specialist chest clinics and over half of whom will require prophylactic therapy. There are therefore at least 3 million sufferers in Britain today. Yet, despite the improvement in treatment available in the past decade there has been a rise in numbers of attacks of acute asthma, notably in children, and a consequent rise in hospital admissions. The same pattern is also seen in adults, albeit less marked. Despite a number of possible incriminating causes, the real reasons for the rise are not clear. Whatever the cause may be, the inevitable effect is clear – the workload of doctors and nurses is expanding with only now, in the mid-1990s, the suggestion that this increase has begun to level off.

Each year around 2000 people die from asthma in England and Wales and two-thirds of these deaths are preventable. The economic costs of asthma are huge. It has been estimated that the direct cost to the NHS (in 1992) is in the order of £340 million per year, £228 million of which goes in prescription costs. In terms of costs to the economy asthma consumes a further £400 million per year, £52 million of which is paid out in invalidity benefit, £5 million in sickness benefits, and the remainder is attributed to lost productivity. Better control of asthma would erode this economic burden on the community.

The British Thoracic Society has proposed guidelines for the management of chronic and of acute severe asthma. An essential part of both these management schemes is the domiciliary use of the peak flow meter, not only to help the patient manage his/her own asthma but also to assist doctors in asthma management.

It is our belief that measurement of peak expiratory flow is fundamental to the management of asthma. It carries an importance equivalent to measuring capillary blood glucose in the management of diabetes. However, in a survey of asthma in general practice published in 1989, 45% of patients with asthma had not even one peak expiratory flow measurement in their records. Until October 1990 peak flow meters were not prescribable in the UK and adequate supplies were difficult to come by largely because of the expense. This barrier to their use has now been removed, leaving the way clear for improvement in asthma control. Unfortunately a survey of asthma referrals to a Respiratory Medicine Department before and after this landmark showed only a slight increase in the use of peak flow monitoring by the referring GP from 22% to 26%.

In asthma substantial airflow obstruction may occur with little in the way of symptoms. As many as 15% of patients with asthma are unable to perceive the presence of marked airways obstruction. In some, airflow obstruction may develop so slowly that the patient begins to feel that it is 'normal' to be short of breath and so may remain untreated for years. One study showed that in a slowly developing attack 19% of asthmatics would allow their asthma to deteriorate for a week before seeking medical help, and 17% of asthmatics would not have sought help despite finding it difficult to speak or to rise from a chair! Failure of perception of the severity of asthma may also be a factor in poor compliance with therapy. In clinical terms this may lead some patients to develop severe degrees of airway obstruction with relatively little discomfort, and mild asthmatics to take inappropriately large amounts of relief inhaler. These problems can be resolved by using an objective measurement, and this is most easily available with a peak flow meter.

As with any measurement, correct use of the meter, appreciation of the benefits and the limitations of the test, and appropriate interpretation of the results are crucial factors to its successful application. The use of peak flow measurement by patients and the adjustment of treatment accordingly, reduces both the number of asthma attacks and the number of oral steroid courses needed. If all patients in the UK with asthma were prescribed peak flow meters, the cost would be £18 million (1992 prices). Appropriate use is therefore mandatory to justify this cost outlay.

The aim of this book is to provide an introduction to peak flow measurement, its indications, its uses, its benefits . . . and its potential pitfalls. We have designed it to be a readable reference book with examples of classic peak flow records and worked case histories. The majority of the examples we have used have been taken from patients we have seen in hospital or general practice. None of the peak flow charts has been fabricated to make a point. We hope the book will be of use to general practitioners and their practice nurses and to hospital clinicians. It is also intended to be of use to medical students, to nursing staff on respiratory and general medical wards and to nurses in schools.

What is peak flow? 1

Shall man into the mystery of breath
From his quick beating pulse a pathway spy?

George Meredith, *Hymn to Colour*

The bronchial tree transmits air at a rate which depends primarily on the overall diameter of the tubes through which the air is passed. The fundamental characteristic of asthma is the tendency for the airways to narrow and widen either spontaneously as a reflection of the disease process or as a result of intervention with smooth muscle relaxing and anti-inflammatory drugs.

1.1 Lung volumes

In a healthy adult at rest, the tidal volume drawn into the lungs at each breath is of the order of 450 ml (Fig. 1.1). The available lung volume is called the vital capacity. It is the difference between total lung capacity at full inflation and residual volume at full expiration. Vital capacity can be measured either by a fast ('forced') or by a slow manoeuvre but in a healthy individual there is no significant difference between the two.

If the airways narrow, for example during an episode of asthma, the rate at which air can be expelled is reduced. Flow ($\dot{V}$) through a rigid tube is proportional to the fourth power of the radius (a) of the tube ($\dot{V} \propto a^4$), so very small changes in tube diameter can have profound effects on flow, for instance a 10% increase in diameter will increase flow by 40%. The bronchial tree is not a rigid tube system but its degree of rigidity is sufficient for this principle to apply.

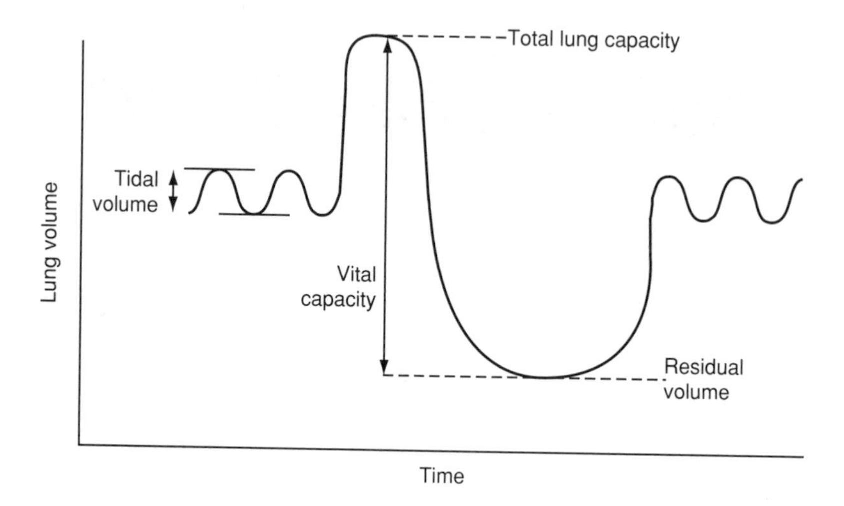

Figure 1.1 Diagrammatic representation of the various lung volumes. After breathing tidally, the subject has breathed in to total lung capacity (TLC) then expelled his vital capacity (VC) by a forced expiratory manoeuvre to reach residual volume (RV). This is followed by inhalation and restoration of tidal breathing.

1.2 Airflow in asthma

Airway narrowing in asthma is due to a combination of three factors:

1 bronchial smooth muscle constriction;
2 mucosal oedema;
3 free mucus secreted into the airways.

Bronchial smooth muscle constriction can cause sudden airway narrowing which may be quickly and easily reversed by use of an inhaled bronchodilator. In more persistent asthma where changes in mucosal oedema and airway mucus contribute more to airway narrowing, changes in flow may not be so sudden. Nevertheless, all these processes will contribute to reduced airflow. In general terms, narrow tubes cause low flows.

1.3 Measurement of flow

There are several ways of expressing flow. In the lung, flow has to be measured at the mouth and therefore reflects the sum of the different resistances to flow from all generations of bronchi (Fig. 1.2). To understand the different measures of flow which can be obtained, it is useful to start with the forced expiratory flow volume curve.

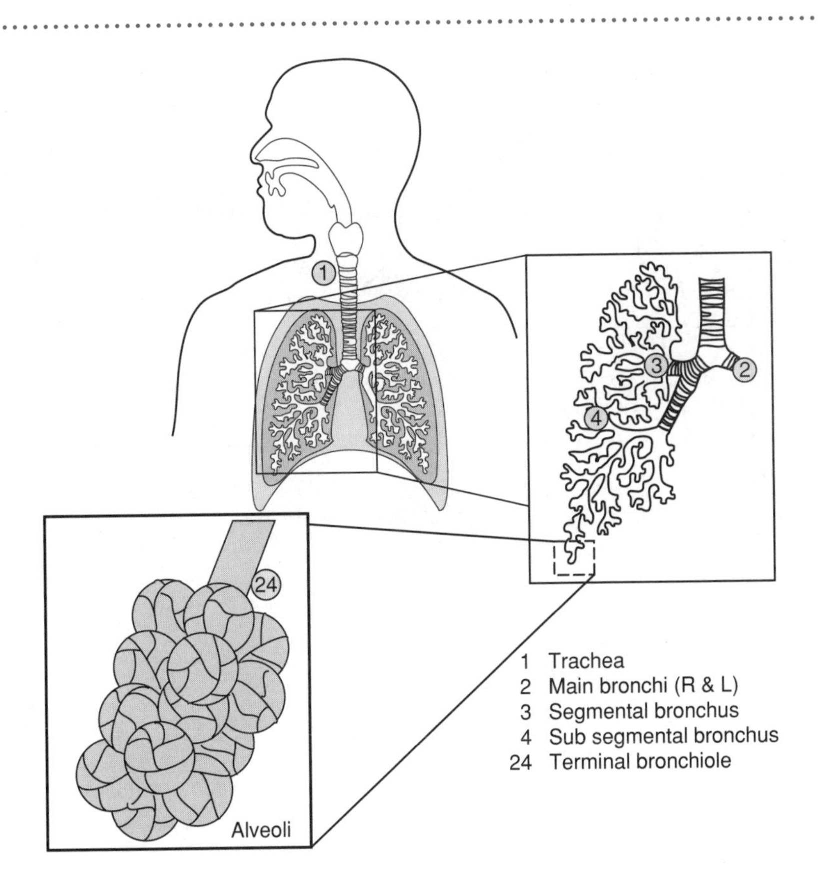

Figure 1.2 The bronchial tree. The airway subdivisions are enumerated starting at No. 1.

To generate a flow volume curve the patient inspires to total lung capacity and then breathes out as hard and as fast as possible to residual volume. The exhaled air is passed through a pneumo-tachograph which gives a flow, and is plotted on the vertical axis, against volume on the horizontal axis (Fig. 1.3). It is important to realize that there is no time axis on a flow volume curve. Peak expiratory flow is reached within 100 msec and is usually defined as the maximum airflow achieved during a forced expiration from total lung capacity. In a normal subject flow then falls in a linear fashion to zero when residual volume is reached. In a patient with airflow obstruction flow does not decline linearly, the curve giving a scooped appearance (Fig. 1.4).

The received wisdom is that flow measured towards the end of a forced expiratory manoeuvre (i.e. at low lung volumes) is limited by the resistance of the small airways, whereas flow towards the beginning of the manoeuvre is limited by large airway resistance. Thus, peak expiratory flow is mainly an index of resistance to flow through the

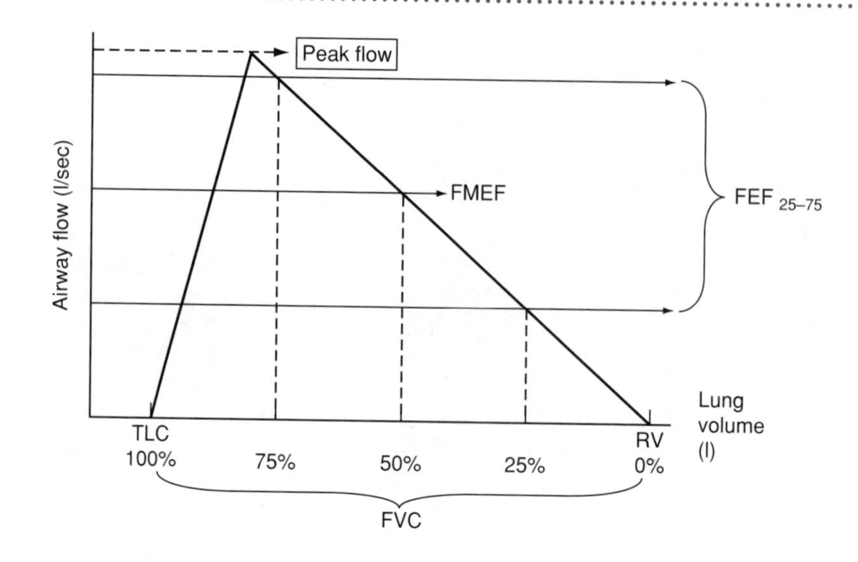

Figure 1.3 From this curve, flows at different lung volumes can be measured, e.g. the forced (or maximum) mid-expiratory flow (FMEF or MMEF) which measures flow at 50% vital capacity. Another measure is the average of flows between 25 and 75% of vital capacity, the forced expiratory flow (FEF_{25-75}). TLC = total lung capacity; RV = residual volume; FVC = forced vital capacity.

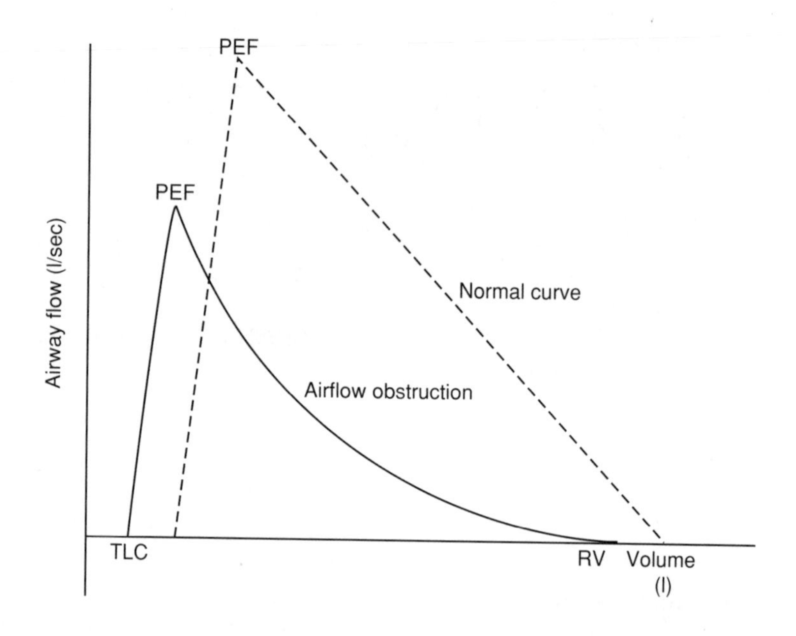

Figure 1.4 Curve in a patient with airflow obstruction. PEF = peak expiratory flow; TLC = total lung capacity; RV = residual volume.

larger airways whereas the other flow measurements from the flow volume curve are better indicators of resistance to flow through small airways. However, in practical terms, the best indicator of small airway resistance to flow is the forced expired volume in one second, or FEV_1 (see below). In clinical, day-to-day practice, flow is measured using a peak flow meter or a spirometer.

1.3.1 Spirometry

Spirometry, using the dry bellows spirometer (Fig. 1.5), has long been regarded as the standard measure of basic lung function. The patient breathes in to total lung capacity and then exhales as fast and for as long as possible into an expanding bellows, which extends a metal arm over a moving chart, thus generating a curve of volume against time (Fig. 1.6). The total volume exhaled is the vital capacity, the amount exhaled in the first second being called the FEV_1. Theoretically one could measure peak expiratory flow from this trace by taking the slope of the trace at the very onset of expiration. In practical terms this is impossible although with some spirometers a reading of peak flow is computed. However, it should be stressed that in a spirometric manoeuvre where the patient is exhaling to residual volume, the measured peak flow tends to be lower than peak flow measured by a peak flow meter, where the forced expiration is terminated quickly.

The value of spirometry is that it can differentiate between restrictive and obstructive defects (Fig. 1.7), which peak flow alone cannot do (see Chapter 3).

In an obstructive defect such as is seen in asthma or chronic obstructive bronchitis, flow is reduced and so PEF and FEV_1 are below predicted values. However, in these conditions the vital capacity may well be normal or be reduced to a much lesser degree than FEV_1. Conse-

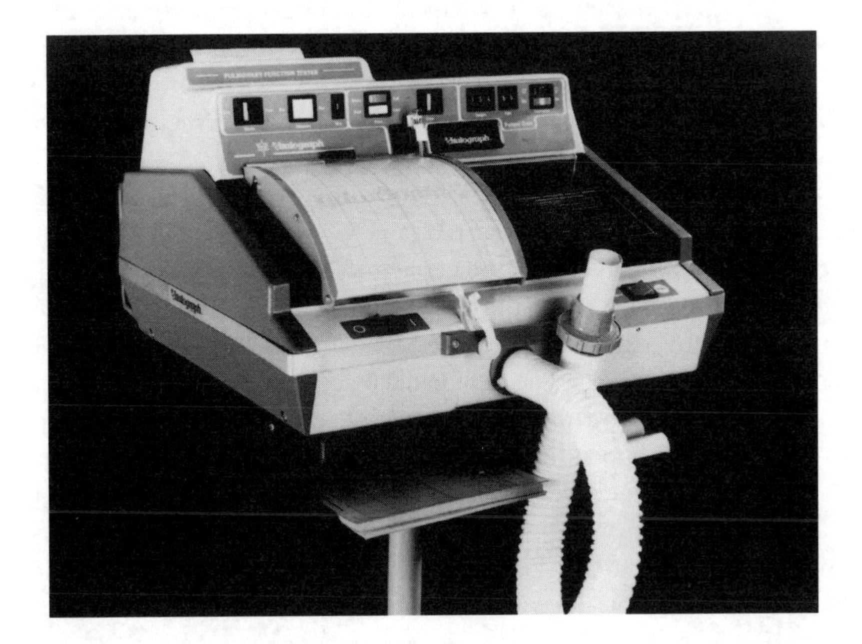

Figure 1.5 The dry bellows spirometer.

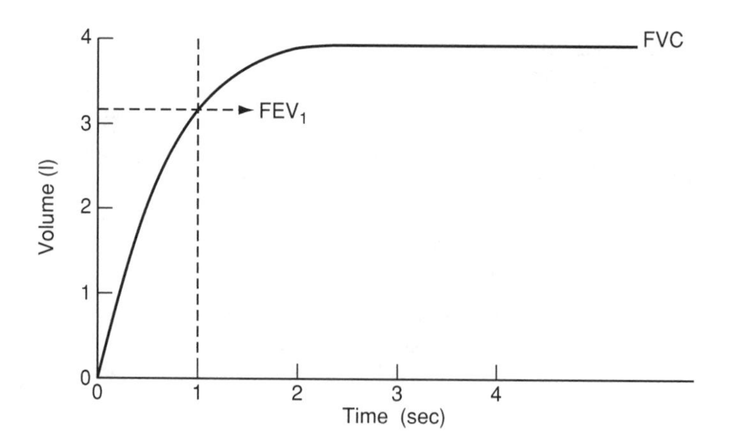

Figure 1.6 Spirometer trace from a normal subject. FVC = forced vital capacity; FEV_1 = forced expiratory volume in one second.

quently calculation of the forced expiratory ratio (FEV_1 divided by FVC%) shows that a smaller percentage of the vital capacity is expelled in the first second than in most normal subjects, who expel 75% in that period.

In contrast, a patient with a restrictive defect (e.g. fibrosing alveolitis or kyphoscoliosis) may have a PEF and FEV_1 similar to a patient with airflow obstruction. Obstructive and restrictive defects can be differentiated as in a restrictive defect the vital capacity is limited to the same degree as FEV_1 so the forced expiratory ratio is normal. Another way of separating an obstructive from a restrictive defect in this situation is to measure the forced expired time (FET), the time taken to exhale to residual volume. In airflow obstruction FET is prolonged, particularly in patients with chronic airflow obstruction, whereas in a restrictive defect FET is normal, i.e. up to 4 seconds.

1.3.2 PEF compared to FEV_1

Peak flow meters are small, portable and enable the patient to take multiple readings during the day. They are also cheap. Conversely spirometers are expensive and the standard bellows spirometers are large and not portable. The new hand-held electronic spirometers are portable and will, in time, enable multiple domiciliary measurements to be made (see Appendix A), but they remain expensive.

Both manoeuvres have in common the need for a forced or maximal effort to generate peak or maximal flow. This requires the effort and co-operation of the patient. Peak expiratory flow is effort-dependent and lack of effort does produce low readings. In contrast FEV_1 is much less

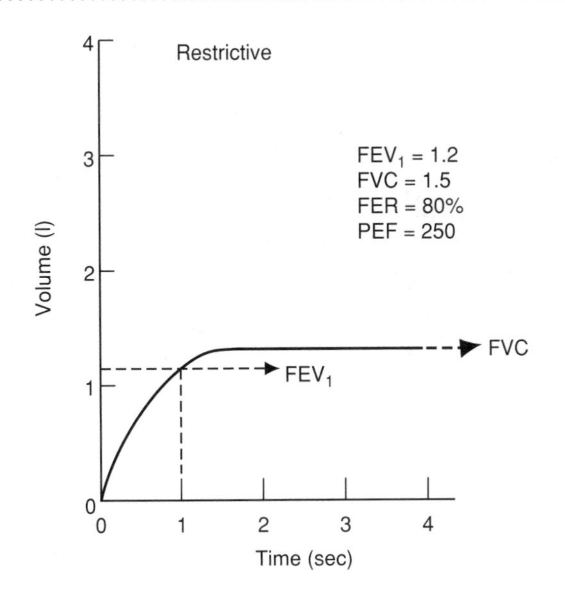

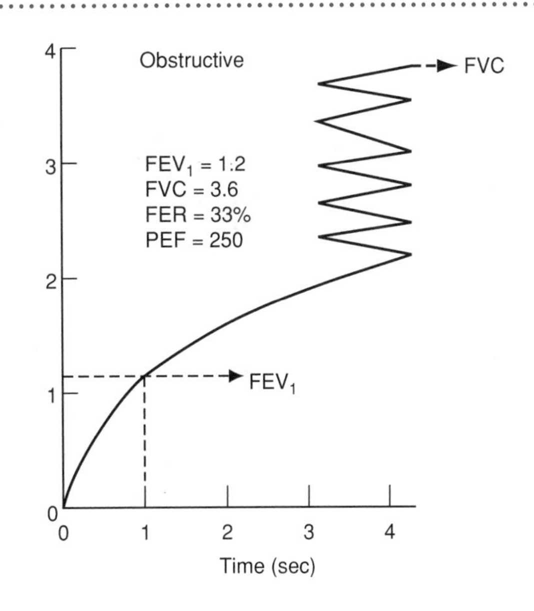

Figure 1.7 Differentiation of a restrictive from an obstructive defect using spirometry. The zig-zag pattern represents alternating movement of the carriage due to the prolonged forced expiratory time (FET) of the patient with airflow obstruction.

effort-dependent, although in some individuals it appears impossible to obtain a reproducible trace. Equally essential is the determination of the technician, physician or nurse to ensure that such reproducible and accurate measurements are obtained.

It is often believed that spirometry remains within the domain of the hospital chest clinic. This need not necessarily be the case and its use could easily be extended into general practice where spirometry can help in differentiating obstructive from restrictive lung problems and in helping to differentiate between asthma and chronic airflow obstruction in its various forms (chronic obstructive bronchitis, chronic productive bronchitis and emphysema).

For instance, Fig. 1.8 shows the peak flow chart of a 62-year-old man with episodic breathlessness who had stopped smoking cigarettes some 15 years previously. The peak flow showed a characteristic diurnal variation with a consistent bronchodilator response, but his spirometry showed a dramatically reduced FEV_1. His flow volume curve and low gas transfer suggested that he had quite marked emphysema, despite his apparently asthmatic peak flow chart. This reinforces the value of spirometry as an adjunct to the screening use of peak flow. Although there is often considerable overlap diagnostically, physiologically and pathologically with asthma and chronic airflow obstruction, it is important to realize that patients with the floppy airways of emphysema where the elastic support of the lung is lost, have this characteristic

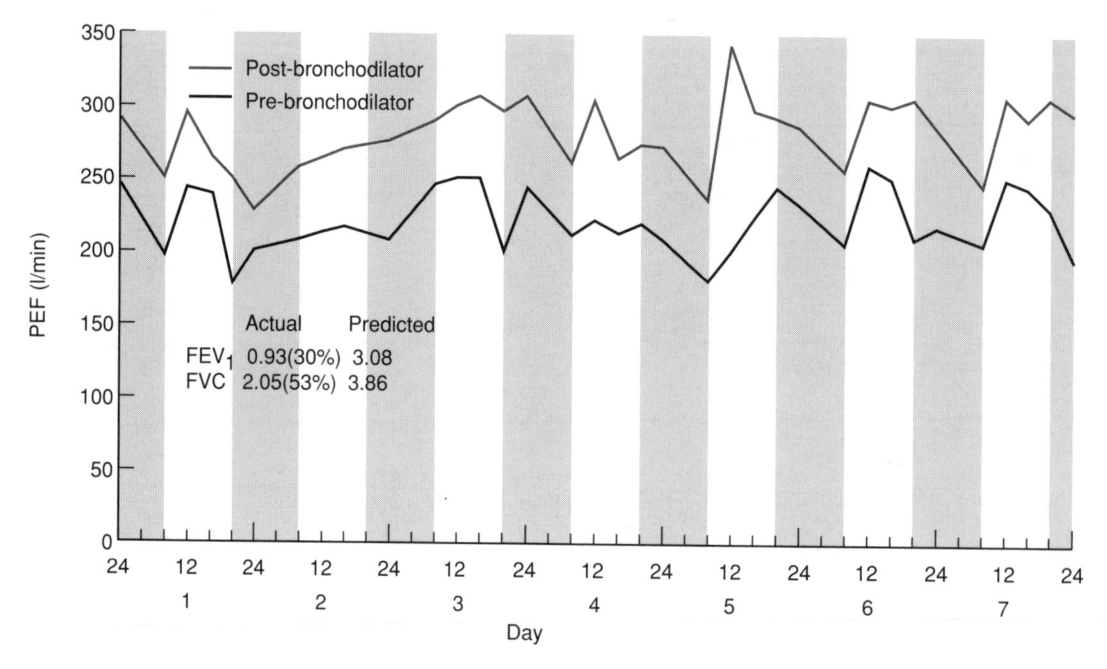

Figure 1.8 Peak flow chart of a 62-year-old ex-smoker with episodic breathlessness. Peak flows show a consistent response to bronchodilators but spirometry shows severe airflow obstruction.

expiratory flow volume curve often with a reasonably well-maintained peak flow, but a low FEV_1. On the other hand, a patient with chronic obstructive bronchitis or asthma shows a more subtle curve (Fig. 1.4), where FEV_1 may be considerably higher than in the patient with emphysema despite the same peak flow.

The advantages and disadvantages of the two manoeuvres are compared in Table 1.1.

Table 1.1 Advantages and disadvantages of peak flow and spirometry

Peak flow	*Spirometry*
Apparatus cheap and portable	Apparatus expensive (some portable)
Manoeuvre simple	Manoeuvre difficult
Only gives a flow	Gives both a flow and a volume
Measures large airway flow only	Measures both large and small airway flow combined
Effort-dependent	FEV_1 effort-independent, but FVC effort-dependent

1.4 Causes of reduction in peak expiratory flow

A reduced peak expiratory flow can be caused by many different diseases and conditions (Table 1.2). Obstructive lesions of the larynx, trachea or main bronchi may result in a low peak flow, but by far the commonest cause of reductions in peak expiratory flow are those due to diffuse airway diseases such as asthma, chronic obstructive bronchitis, chronic productive bronchitis and bronchiectasis. However, conditions causing lung restriction can also cause reductions in peak flow and here differentiation between airflow obstruction and lung restriction will be made by spirometry (see section 1.3.2), where peak flow alone will not define the problem. Lung restriction can be caused by interstitial lung disease such as fibrosing alveolitis, pleural disease (such as post-tuberculous fibrosis or fibrosing mediastinitis), or disorders of the thoracic cage such as kyphoscoliosis, post-surgery or ankylosing spondylitis.

Table 1.2 Causes of reduced peak flow

Larynx	tumour
	laryngeal oedema
	laryngeal spasm
Trachea	tumours
	goitre
	strictures (e.g. post endotracheal tube)
	relapsing polychondritis
Bronchi	tumour
	inhaled foreign body
	widespread airway disease
	asthma
	chronic obstructive bronchitis
	chronic productive bronchitis
	emphysema
	bronchiectasis
Lung parenchyma	interstitial fibrosis*
	(e.g. cryptogenic fibrosing alveolitis)
Pleura	tumours (e.g. mesothelioma)*
	fibrosis (e.g. post tuberculosis)*
Chest wall	scoliosis*
	ankylosing spondylitis*
	neuromuscular disease
Heart	'cardiac asthma'

*These cause a restrictive defect (see text).

1.5 Factors governing peak flow

For details of the history of peak flow measurements and available types of meter, the reader is referred to Appendix A.

Peak flow is dependent on three main factors: height, age and sex. In children peak flow is related to height (Fig. 1.9) but not to sex, although a clear sex difference develops in the teenage years. Peak flow reaches a maximum between the ages of 30 and 35 (Fig. 1.10) and then declines gradually in common with other lung function parameters.

There have been a number of normal ranges published for peak expiratory flow in white populations, for example those produced by Cotes, by Gregg and Nunn (obtained from non-smokers), and by the European Commission for Coal and Steel (ECCS) (giving slightly lower normal values obtained from a group containing some smokers). In

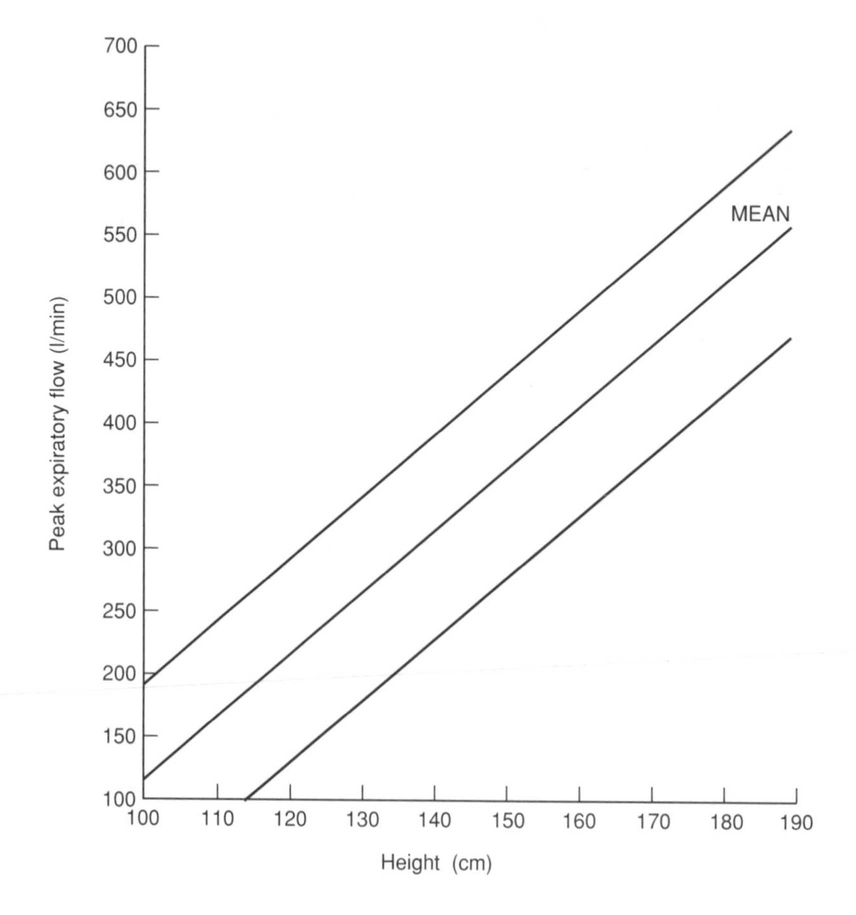

1.9 Relationship of ... peak flow in children. ...n from original data of ... *et al.* (1970), *British* ... *of Diseases of the Chest*, **64**, 15.)

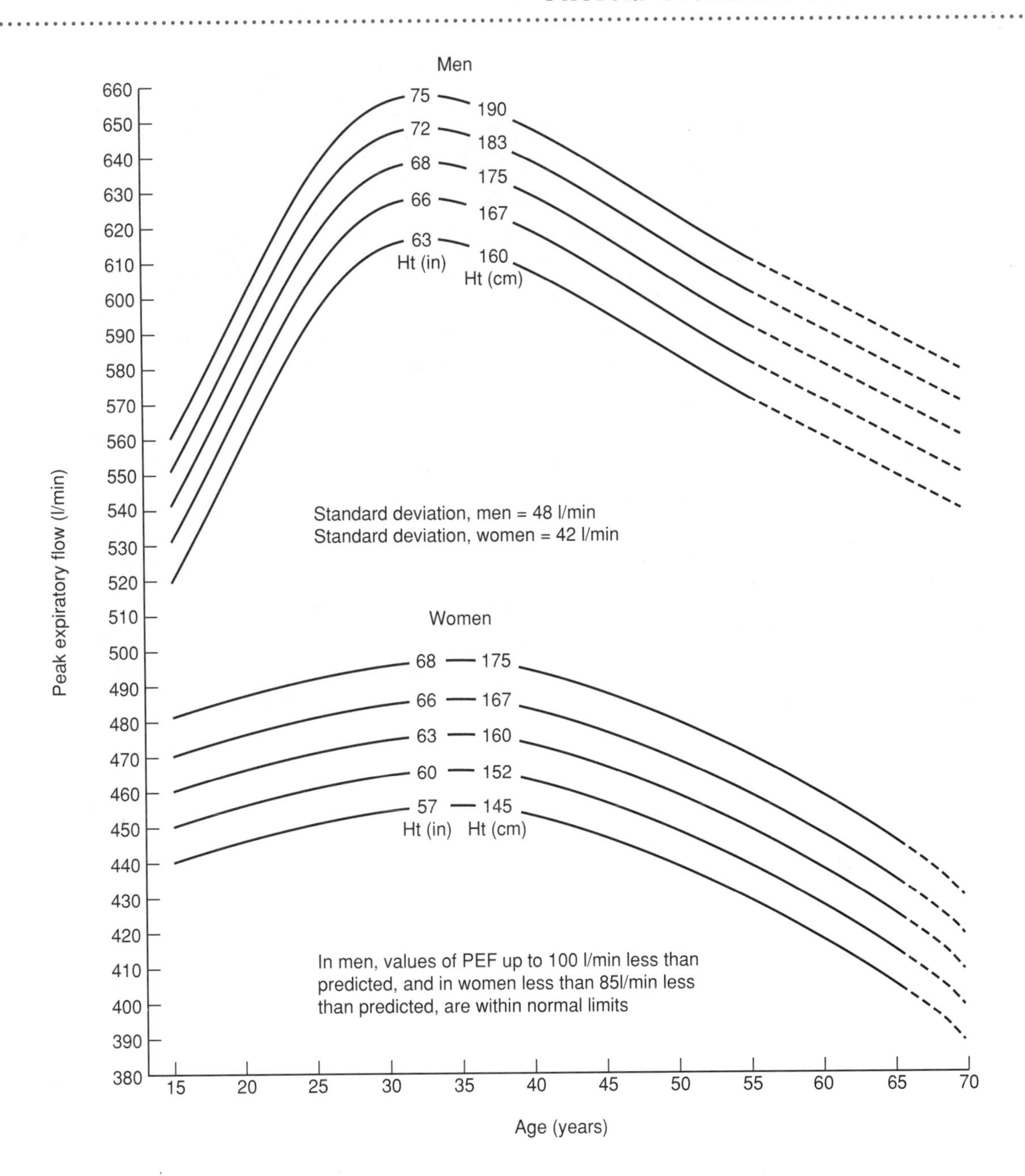

Figure 1.10 Peak expiratory flow in normal subjects. (Redrawn from Gregg and Nunn (1973), *British Medical Journal*, **iii**, 282 and reproduced with permission.)

Europe the ECCS set of values is coming into more general use. The ECCS tables are given in Appendix B.

It is important when considering any indicator of lung function that uniformity of normal ranges is used. Comparisons can then be made between separate published studies. In the clinical field a change from one normal range to another may have significant implications if the data are expressed in terms of percentage or predicted value. Recent concerns about the non-linearity of peak flow meters are discussed and put into context in Appendix A.

There is a degree of variation in normal ranges of peak expiratory flow between different ethnic groups and series of lung function tables have been published of many different ethnic groups both in their land of origin and after having moved to foreign lands. In the past it was usual to reduce the predicted values by 10% for adult patients of Afro-Caribbean or Indian subcontinental origin, but some feel that this is no longer necessary.

Formulae dependent on height, age and sex can be used for prediction of peak flow in an individual but these are usually only of use in research projects of an epidemiological nature. Nomograms have also been used but the margin for error in their use is relatively small and it is probably best to stick to tables (see section 2.8).

Measurement, recording and analysis of peak flow records

2

'I can't believe that!' said Alice. 'Can't you?' the Queen said in a pitying tone. 'Try again: draw a long breath, and shut your eyes.'

Lewis Carroll, *Through the Looking Glass*, ch. 5

2.1 How to measure peak flow

The technique of measuring peak flow is simple but, like inhaler technique, must be taught and rechecked at later consultations.

A careful explanation, followed by a demonstration by the doctor or nurse, is essential and helpful.

2.1.1 Reproducibility

Steps for obtaining peak flow

1 Check pointer is at zero (Fig. 2.1).
2 Hold meter horizontally.
3 Inhale fully.
4 Pause slightly, then put mouthpiece into mouth ensuring teeth are around the mouthpiece and that there is an airtight seal around mouthpiece.
5 Blow out as hard and fast as possible and avoid puffing out cheeks.
6 Read off value (Fig. 2.2).

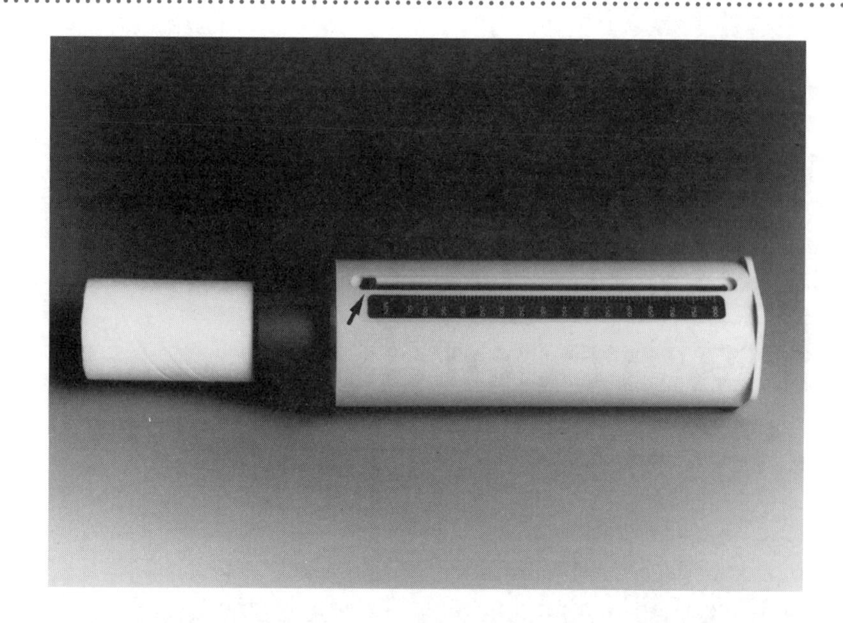

Figure 2.1 Mini Wright peak flow meter. Arrow shows pointer at zero.

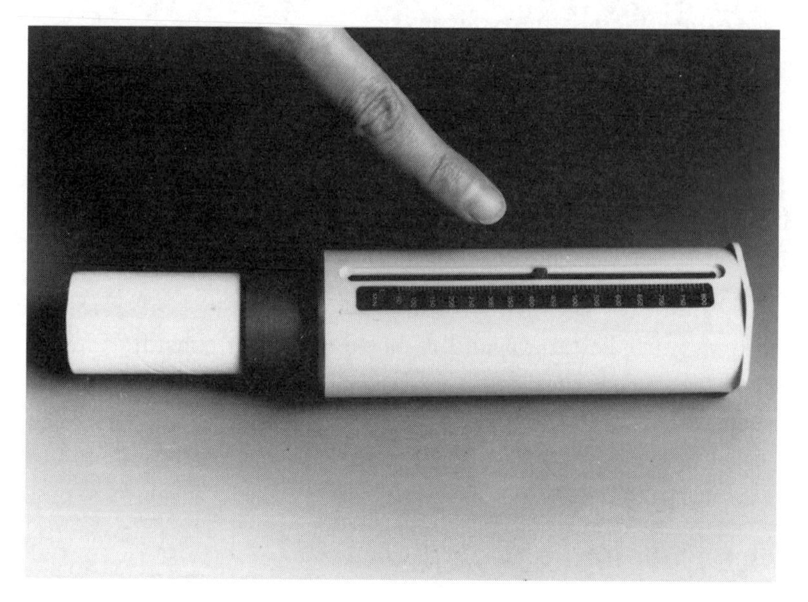

Figure 2.2 Position of pointer after an expiration.

7 Repeat manoeuvre two or more times until three readings are within 20 l/min of each other.
8 Select and record the best of the three readings.

Generally the highest of three recordings is taken as the peak flow at a particular time. Most children can use a peak flow meter effectively from the age of 5 years, but five blows may be needed to get the most representative value in children. A path must be steered between getting the best peak flow and the child becoming bored. However with successive blows children seem to concentrate more and improve their technique.

2.1.2 Pitfalls in measuring peak flow

Measurement of peak flow may be subject to error for a variety of reasons. These are almost all avoidable with adequate coaching in the use of the meter at the outset.

Peak flow is to a large extent dependent on effort, and by necessity it is often measured at home without an assessment of the effort put in or the accuracy of recordings. In fact most patients keep reliable records. Readings may, however, be erroneous due to obstruction of flow by the tongue or by poorly fitting dentures, or due to obstruction of the cursor by a finger on the scale. There must be an adequate rest period between successive attempts at measuring peak flow. In some patients with marked bronchial reactivity the test itself can lead to bronchoconstriction and hence to much lower values at the second and third attempts if peak flows are taken in quick succession. Falsely high readings may be obtained by a spitting action using the tongue and cheek muscles rather like a trumpet (Fig. 2.3). This problem can be avoided by ensuring that the mouthpiece is well inside the mouth before blowing.

Pitfalls in measurement are summarized in Table 2.1.

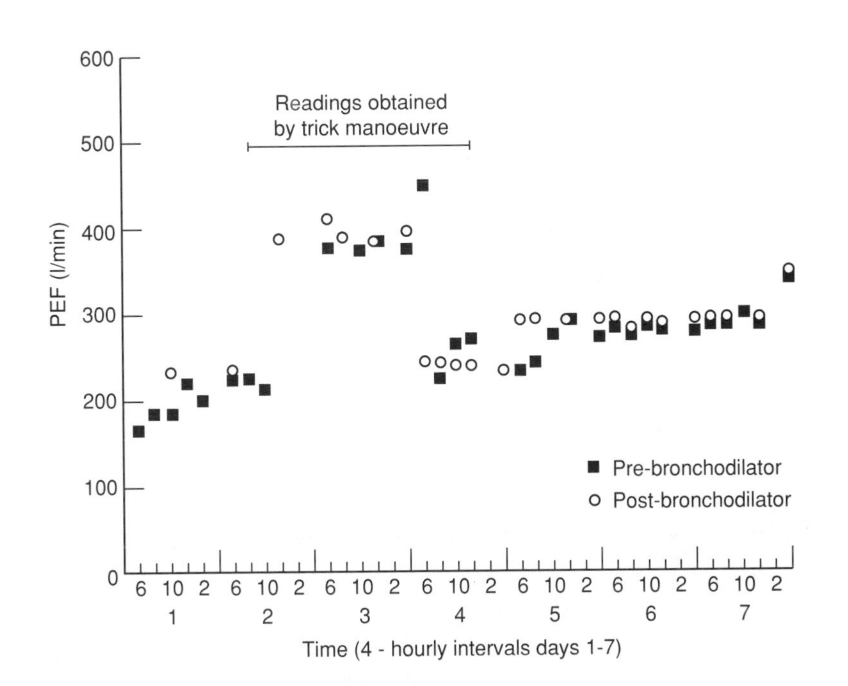

Figure 2.3 Higher values of peak flow obtained by a 'spitting' action using the tongue and cheek muscles like a trumpet. (Redrawn from C.K. Connolly (1987), *British Medical Journal*, **2**, 285 and reproduced with permission.)

Table 2.1 Pitfalls in measurement

1	**Poor effort**
	Confusion
	Weakness
	Lack of motivation
	Lack of coordination
2	**Errors in use of mouthpiece**
	Leaks around mouthpiece
	Lips not round mouthpiece
	Facial palsy
	Obstruction by tongue
	Poorly fitting dentures
	'Spitting action' of cheek muscles and tongue leading to falsely high readings
3	**Obstruction of cursor**
	Fingers
4	**Obstruction of air vents**
	Fingers
5	**Inadequate rest period between successive peak flow attempts**
6	**Deliberate falsification**

2.2 When to measure peak flow

Acute asthma Peak flow is reduced in acute asthma and is one of the major indices of its severity. It must be measured in every asthma attack either by the patient or by the attending doctor. This gives an indication of the severity of an attack when taken with other factors such as tachycardia, central cyanosis, low PO_2 and a silent chest. A response to therapy, or lack of it is then readily apparent. In an acute attack peak flow should be measured at least four times daily both before and ten minutes after a bronchodilator. This frequency of measurement should continue until the peak flow has returned to optimal levels with little diurnal variation (section 3.4, Acute asthma). Patients should be advised to measure peak flow at other times if they detect a deterioration, such as being awoken by asthma at night; they should not just measure peak flow strictly four times per day.

Unstable asthma This is recognized by increasing use of bronchodilators, nocturnal wheeze and morning dipping (described below). Guidelines for action at particular levels of peak flow can be tailored to the individual patient, who should be aware that falling peak flow or increased diurnal variation are both situations requiring action, to avoid crises (Chapter 5, Management plans).

Chronic asthma Diurnal variation is a measure of the effectiveness of prophylactic medication. A patient's predicted peak flow for age, height and sex, **or their maximum achievable value after a trial of oral corticosteroids**, becomes the target for treatment with inhalers. Serial peak flows will record small but significant changes in their asthma that would otherwise be missed.

Well-controlled asthma The frequency of measurement will depend on individual patients, and often measurement may only be needed once daily. These patients will have been identified as well controlled when previously peak flows were monitored more frequently.

2.2.1 Multiple versus single measurements

Single measurements of peak flow in the clinic or surgery can be very misleading and can lead to a false sense of security. The story that peak flow readings tell depends largely on the frequency of recordings. The four times daily peak flow chart shown at the bottom of Fig. 2.4 displays the chaotic pattern of 'brittle' asthma described in section 3.5 below, but the middle recording shows the same patient with measurements at 8 am and 8 pm only. He appears more stable, despite his report of recurrent wheeze! Even more reassuring, but misleadingly so are the measurements shown at the top, taken at 4 pm each day, as might come from regular measurements at evening surgery.

2.3 How to record peak flow

If patients are to be asked to record peak flow on charts they should be told to use a pen (not pencil) and use blue or black ink for values recorded before bronchodilator use. Values recorded after a bronchodilator should be recorded in red and written directly above the pre-treatment value on the graph. With children it may well be helpful for the child to select their own colour coding, so long as they don't keep changing!

There are many peak flow diary cards available from drug companies and manufacturers of peak flow devices.

Most charts are for twice or three times daily recording. Among the best are those supplied by Action Asthma with both adult (Fig. 2.5)

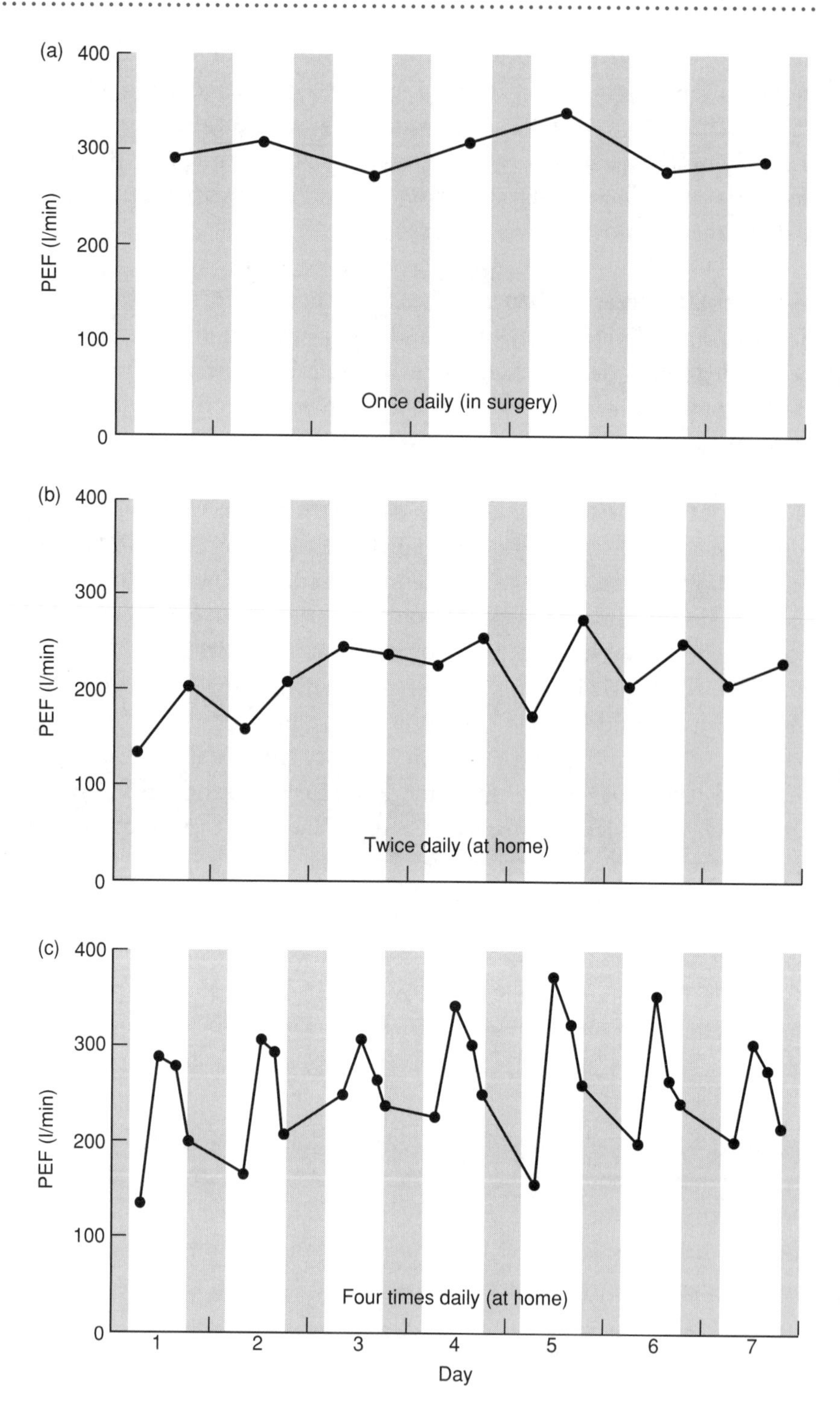

Figure 2.4 The peak flow chart of a patient who was found to have brittle asthma. Pre-bronchodilator values have been used. (*a*) Single daily values recorded at the time of an evening surgery, showing relatively little variability; (*b*) twice daily readings (on waking and on going to bed); (*c*) four times daily readings.

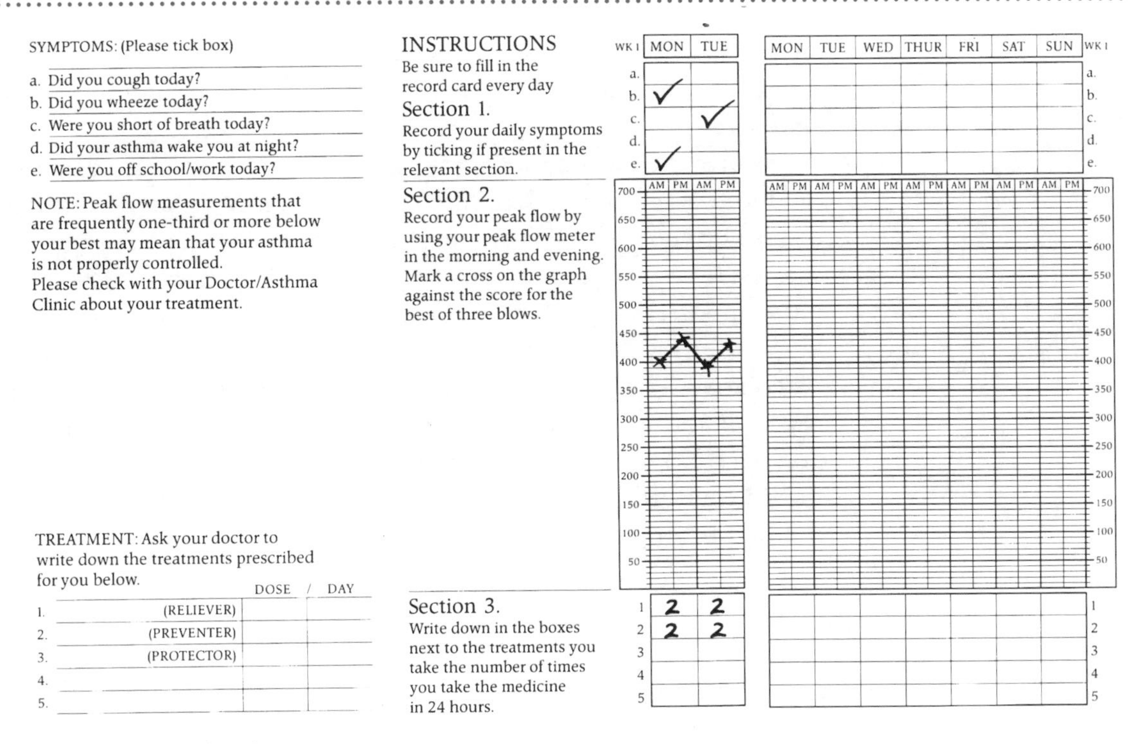

SYMPTOMS: (Please tick box)

a. Did you cough today?
b. Did you wheeze today?
c. Were you short of breath today?
d. Did your asthma wake you at night?
e. Were you off school/work today?

NOTE: Peak flow measurements that are frequently one-third or more below your best may mean that your asthma is not properly controlled.
Please check with your Doctor/Asthma Clinic about your treatment.

TREATMENT: Ask your doctor to write down the treatments prescribed for you below.

		DOSE / DAY
1.	(RELIEVER)	
2.	(PREVENTER)	
3.	(PROTECTOR)	
4.		
5.		

INSTRUCTIONS
Be sure to fill in the record card every day

Section 1.
Record your daily symptoms by ticking if present in the relevant section.

Section 2.
Record your peak flow by using your peak flow meter in the morning and evening. Mark a cross on the graph against the score for the best of three blows.

Section 3.
Write down in the boxes next to the treatments you take the number of times you take the medicine in 24 hours.

WK 1	MON	TUE
a.		
b.	✓	
c.		✓
d.		
e.	✓	

WK 1: MON TUE WED THUR FRI SAT SUN — a. b. c. d. e.

AM PM AM PM … — 700 650 600 550 500 450 400 350 300 250 200 150 100 50

	MON	TUE
1	2	2
2	2	2
3		
4		
5		

Figure 2.5 Adult record card.

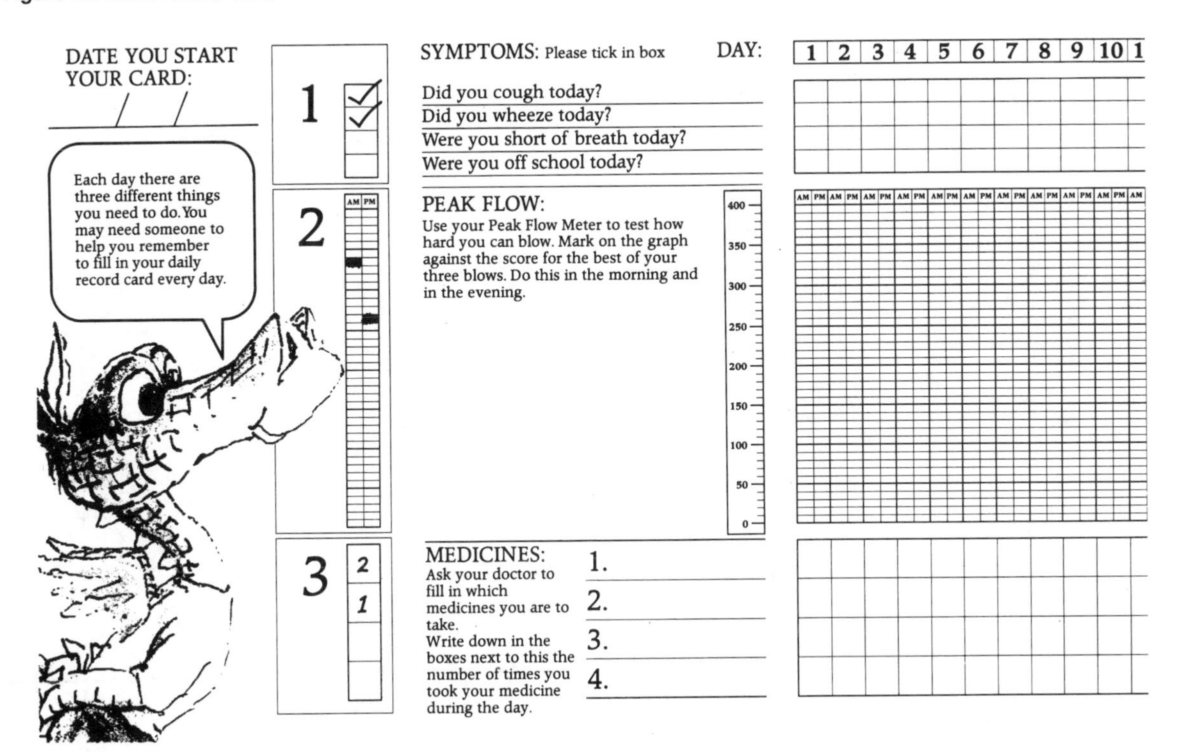

DATE YOU START YOUR CARD: / /

1 2 3

SYMPTOMS: Please tick in box

DAY: 1 2 3 4 5 6 7 8 9 10 1

Did you cough today?
Did you wheeze today?
Were you short of breath today?
Were you off school today?

PEAK FLOW:
Use your Peak Flow Meter to test how hard you can blow. Mark on the graph against the score for the best of your three blows. Do this in the morning and in the evening.

400 350 300 250 200 150 100 50 0

MEDICINES:
Ask your doctor to fill in which medicines you are to take.
Write down in the boxes next to this the number of times you took your medicine during the day.

1.
2.
3.
4.

Figure 2.6 Child's record card.

and paediatric ('Desmond Dragon') (Fig. 2.6) versions. They include space for symptoms and for treatment taken in addition to peak flow values.

For more frequent peak flow recordings, weekly sheets can be used to produce a recording such as that shown in Fig. 2.5 (*bottom*), but these do take up considerable room. It is sometimes helpful, in these circumstances, to ask the patient to retain their peak flow charts, allowing the nurse or doctor to abstract specific details from the charts for their own notes which will not then get weighed down by reams of paper! Alternatively doctors or nurses can devise their own chart on graph paper. If the patient is unable to fill in graphs, or if it is thought necessary to reduce the volume of charts, recording peak flow at particular times in columns (Fig. 2.7) is simple and effective. The more complex occupational peak flow chart is discussed in Chapter 4. In the

Figure 2.7 Simple chart for recording twice-daily measurements.

Date	am	pm	Comments		
01-1-92	320	320	Cough	Ventolin x3	Pulmicort x3
02-1-92	270	290	"	"	"
03-1-92	270	280	"	"	"
04-1-92	270	270	"	"	"
05-1-92	290 250	290	"	"	"
06-1-92	290	300	"	"	"
07-1-92	300	290	"	"	"
08-1-92	290	290	"	"	"
09-1-92	290	310	"	"	"
10-1-92	320	330	-	"	"
11-1-92	330	330	-	-	"
12-1-92	340	350	-	-	"
13-1-92	330	340	-	-	"
14-1-92	330	320	-	-	"
15-1-92	320	330	-	-	-
16-1-92	320	320	-	-	-
17-1-92	330	320	-	-	-
18-1-92	320	330	-	-	-

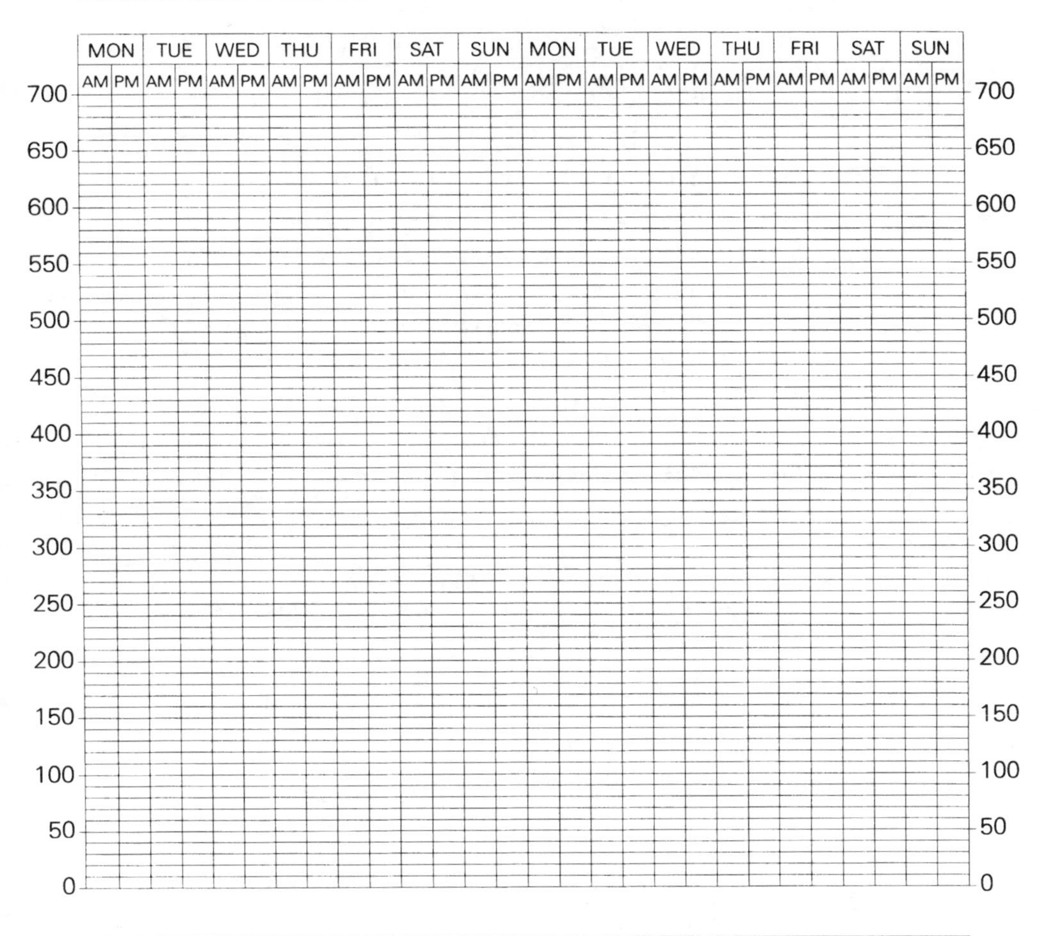
DAILY READINGS
Best of 3 blows, morning and evening.

Date chart started: ______________

	MON		TUE		WED		THU		FRI		SAT		SUN		MON		TUE		WED		THU		FRI		SAT		SUN		
	AM	PM	AM	PM	AM	PM	AM	PM	AM	PM	AM	PM	AM	PM	AM	PM	AM	PM	AM	PM	AM	PM	AM	PM	AM	PM	AM	PM	
700																													700
650																													650
600																													600
550																													550
500																													500
450																													450
400																													400
350																													350
300																													300
250																													250
200																													200
150																													150
100																													100
50																													50
0																													0

	Number of doses of reliever medicine taken to relieve symptoms in 24 hours							Number of doses of reliever medicine taken to relieve symptoms in 24 hours						
	MON	TUE	WED	THU	FRI	SAT	SUN	MON	TUE	WED	THU	FRI	SAT	SUN
Day														
Night														

Comments/Notes : (You should especially record here night-time symptoms or any event such as a cold which may affect readings)

Figure 2.8 Department of Health peak flow chart (Form FP1010).

UK the peak flow charts provided in a book by the Department of Health are good. The charts are prescribed on form FP1010 supplied by the Family Health Service Authority or can be obtained from community pharmacists (Fig. 2.8).

With the advent of intelligent peak flow meters and the more widespread use of computerized asthma management packages, storage problems will become less of a worry.

2.4 Variability in peak flow

The absolute value of peak flow may not be as important as its variability with time. Patients with asthma usually show a diurnal variation in peak flow in excess of 15%, although not necessarily every day, and it is this degree of variability which can be crucial in the diagnosis of asthma and subsequently in assessing response to treatment.

2.4.1 Diurnal variation

Diurnal variation can be calculated in different ways, the calculation varying from worker to worker and country to country. In the UK the usual calculation is:

$$\frac{(\text{daily maximum} - \text{daily minimum})}{\text{daily maximum}}\% \quad \text{i.e. amplitude \% maximum}$$

There is evidence, at least in population studies, that

$$\frac{(\text{daily maximum} - \text{daily minimum})}{\text{daily mean}}\% \quad \text{i.e. amplitude \% mean}$$

may be a more reproducible measure of variability.

However, throughout this book we will use amplitude as a percentage of maximum as an indicator of diurnal variation.

There is some evidence that diurnal variation relates to age in non-asthmatic individuals, children showing a greater range of variation than non-asthmatic adults. Table 2.2 shows the potential range of changes in diurnal variation in peak flow seen in normality. In adults it is generally reckoned that diurnal variation is around 8% in non-asthmatics, and asthma is usually accepted as being present if the diurnal variation exceeds 15%. There is no clear evidence as to whether exceeding a diurnal variation of 15% on just one day would be diagnostic of asthma. Such a finding would need to be taken into consideration along with the clinical picture. If the patient had symp-

Table 2.2 Range of diurnal variation in normal subjects*

Age group (yr)		*Percentiles*		
		95%	*97.5%*	*Median*
6–14	Max/min%	130	140	105.0
	Amp/mean%	31	38	8.2
15–34	Max/min%	117	112	102.7
	Amp/mean%	19	27	5.0
35–65	Max/min%	118	126	103.4
	Amp/mean%	19	29	4.9

* Subjects were non-smokers with no chronic respiratory disease of any sort (including asthma).
Amp/mean% = (max – min/daily mean)%
Source: Quackenboss, J.J., Lebowitz, M.D., Krzyzanowski, M. *et al.* (1991) *Am. Rev. Respir. Dis.*, **143**, 323–330.

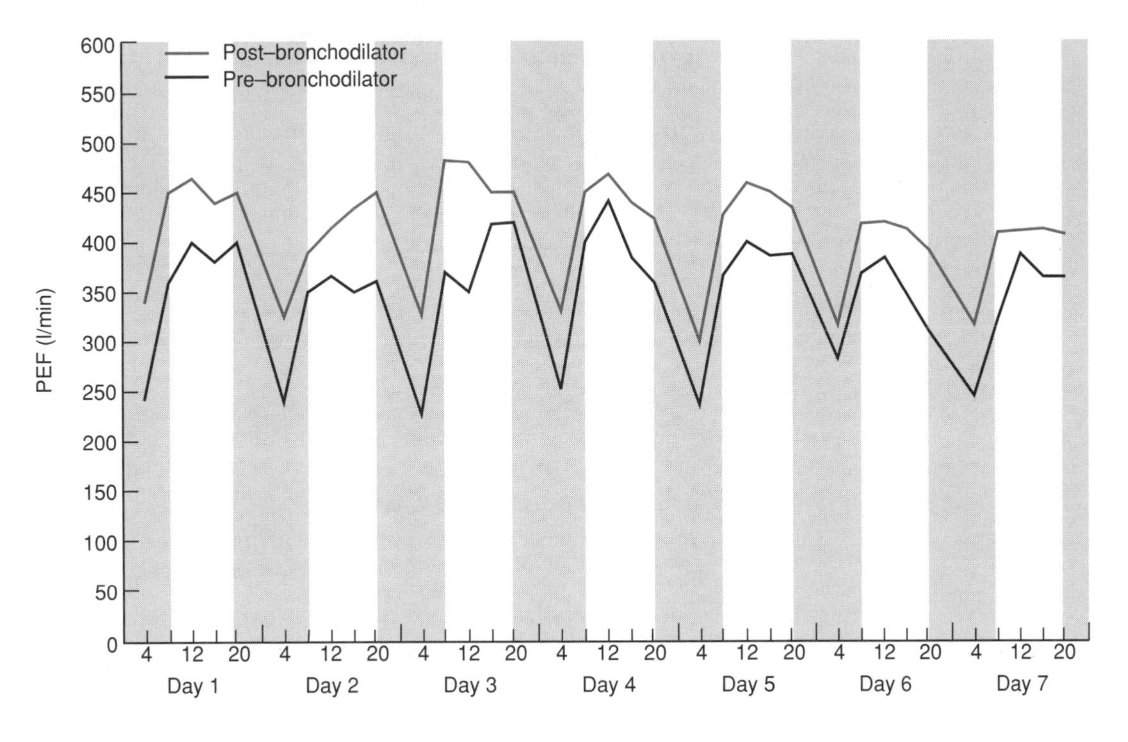

Figure 2.9 Four-hourly recording showing 'morning dipping'.

toms on the day on which the diurnal variation was wide, then this would tend to support a diagnosis of asthma. For this reason, when screening a patient for asthma it is reasonable to ask for domiciliary peak flow monitoring over a long enough period to allow symptoms to occur (section 3.1). A simple table for quick calculation of diurnal variation in peak flow is given in Appendix C.

The most widely recognized diurnal pattern in asthma is 'morning dipping' (Fig. 2.9), where the lowest value is recorded on waking. Another pattern commonly seen is the morning and evening dip. These patterns do not necessarily reflect different types of asthma although patients with profound morning dipping are probably more likely to have nocturnal symptoms implying poor asthma control.

Why this diurnal variation should occur is not known. It has been suggested that in asthma the normal diurnal variation in peak flow is exaggerated and that the shape of the peak flow profile over a 24 hour period is sinusoid and similar to other biorhythms, but this view is not universally held. Indeed, there are many patients with asthma whose diurnal variation in peak flow does not fit such a neat sinusoidal pattern, some dipping in both morning and evening, others showing quite erratic variation. If it is an exaggerated biorhythm then one might expect that correcting the physiological variations in hormones which may have an effect on airway tone, such as cortisol or adrenaline, might abolish this diurnal variation, but this has not been shown to be the case. However, physiological doses of anti-cholinergic agents have reduced such diurnal variation suggesting that cholinergic tone may be important in this regard.

2.5 How to summarize peak flow records

In many cases it will be impossible to summarize a peak flow record in a simple way. In this situation (e.g. initial assessment of a new patient whose chart shows considerable variability), a visual assessment before and after institution of treatment may be all that is necessary.

Peak flow charts can occupy a great deal of room in a patient's clinical record. Asking patients to retain their records (which can then be assessed by all health carers) is a good option, and is more acceptable if the records are in booklet form. However, abbreviations are often needed for inclusion in the patient's notes.

Daily values Day-to-day variation can be expressed in terms of diurnal variation (see above) or as maximum and minimum values, with or without a daily mean value. The latter method is used in assessment of occupational asthma (section 4.2.1), which requires more detailed analysis.

December	260–310
January	260–310 (more chesty, Becloforte to iv b.d.)
February	290–320 (Becloforte to ii b.d.)
March	300–320 (1 value at 250, responded to salbutamol)
April	300–330
May	280–310 (briefly increased Becloforte)
June	300–330

Figure 2.10 Monthly summary of peak flow readings.

Weekly values Again, weekly maxima, minima and means can be used which will express a week's data in two or three readings which can be better stored in the patient's clinical record. It can be helpful sometimes to record the weekly maximum and the usual weekly minimum, noting that on one day a lower value was found. This becomes more useful in patients whose asthma has become more stable. It may only be necessary to record the fact that values were consistently above the target peak (section 5.1). The same approach can be applied to monthly data (Fig. 2.10).

How often peak flow readings are recorded depends on the clinical situation, e.g. two-hourly for occupational asthma, four times daily for new patients or severe asthma and twice daily for milder patients. Recording peak flow is a time-consuming discipline and requires considerable motivation by the patient. Once patients have been assessed then the frequency of measurement needs to be reviewed. Eventually once-daily monitoring by the patient, in the context of a management plan, may be all that is required and some patients may be better and more logically controlled using symptoms alone as a guide to altering therapy (section 5.3).

2.6 Compliance with measurement

Recording peak flow repeatedly over a long period is tedious and it is hardly surprising that some records have missing values. Indeed it is an unusual record that is presented complete! If only a small number of values are missing then this doesn't really affect interpretation of the record, but as the percentage of potential readings that are recorded falls, the more difficult is the task, especially as it is often the waking values that are missed in the rush of the morning.

It is important to let the patient know that you, as doctor or nurse, realize the difficulties but, equally, do not encourage them to feel missing out values 'will be all right'.

Twice-daily peak flows are more likely to be associated with better compliance and it is as well to try to get the patient into a routine, perhaps leaving the meter beside their toothbrush, so that as full a record as can be is obtained.

Nevertheless, everyone is human and missing values will occur, something which some patients feel concerned about and, in order not to 'let the doctor down' fill in values which they feel would have been roughly right. This needs to be discouraged without penalizing the patient, who was not recording false values for perceived gain.

2.7 Falsification of peak flow readings

Concern is often expressed at deliberate falsification of peak flow readings by patients. In practice this is believed to be infrequent although there is no foolproof way of detecting cheating.

There are a few clues which may help in detection of falsification of records, however.

Neatness of records Some peak flow records (either graphical or numerical) appear very neat and pristine giving the appearance of being filled out the night before clinic or surgery attendance. In most cases this will be a genuine rewrite of values either recorded on other sheets or in diaries, but occasionally this should alert the physician to the possibility of falsification. The use of different pens over a period will suggest that the readings are genuine.

Invariability of peak flow values In a patient with symptoms, a straight line of peak flow readings below predicted levels is suspicious. Even in a patient without symptoms (but known to have asthma) such uniformity of readings would be surprising but may reflect the patient's wish to reinforce the message to the physician or nurse that their asthma is well controlled. We have had one patient whose peak flow readings are a straight line when 'well' and then when her asthma 'worsens', becomes suddenly and wildly chaotic, with maximal values well exceeding those when 'well' (Figs 2.11, 2.12).

Unreal variability A problem in occupational asthma, or where environmental factors are thought to cause worsening of a patient's

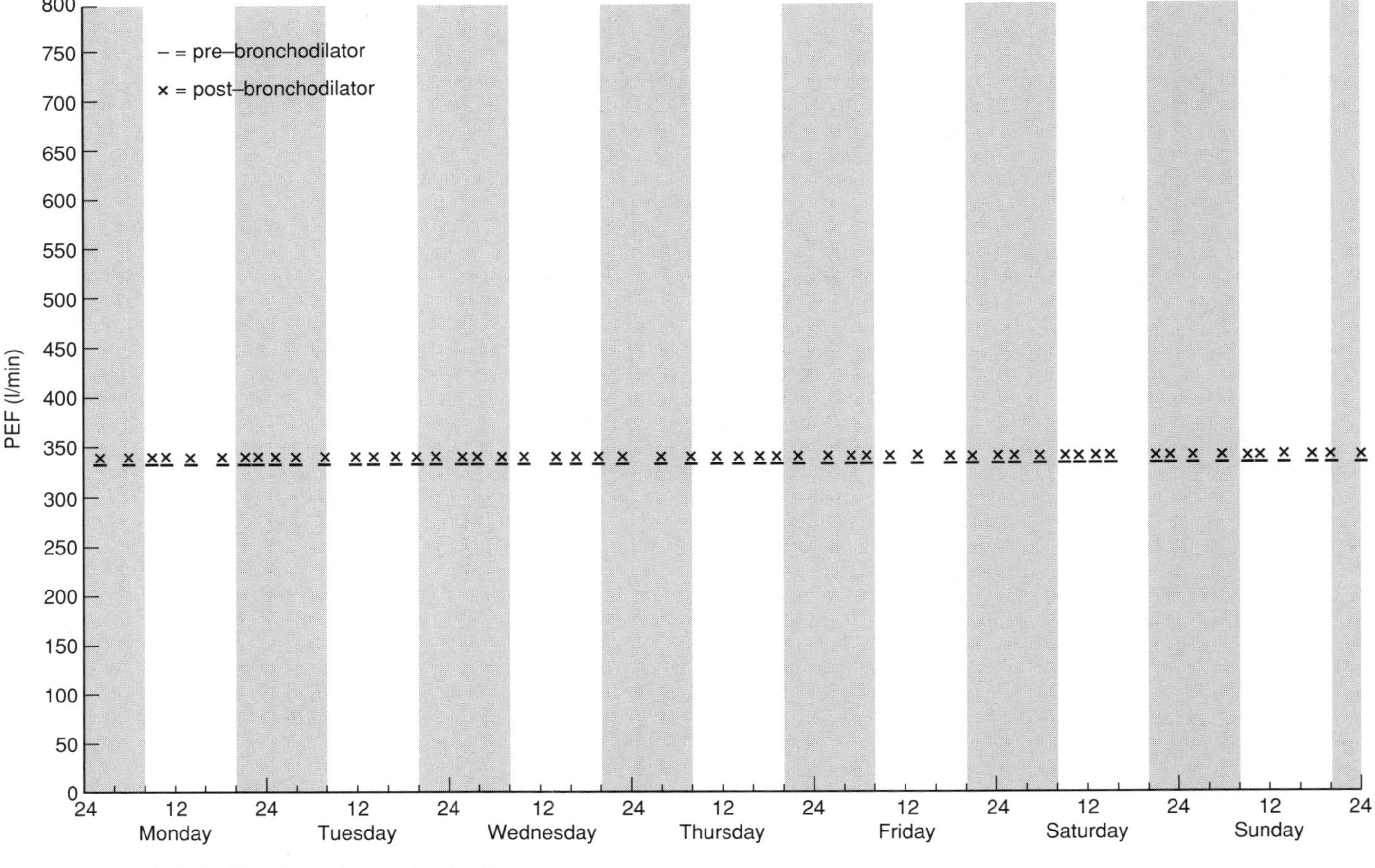

Figure 2.11 Falsified PEF values when patient 'well'.

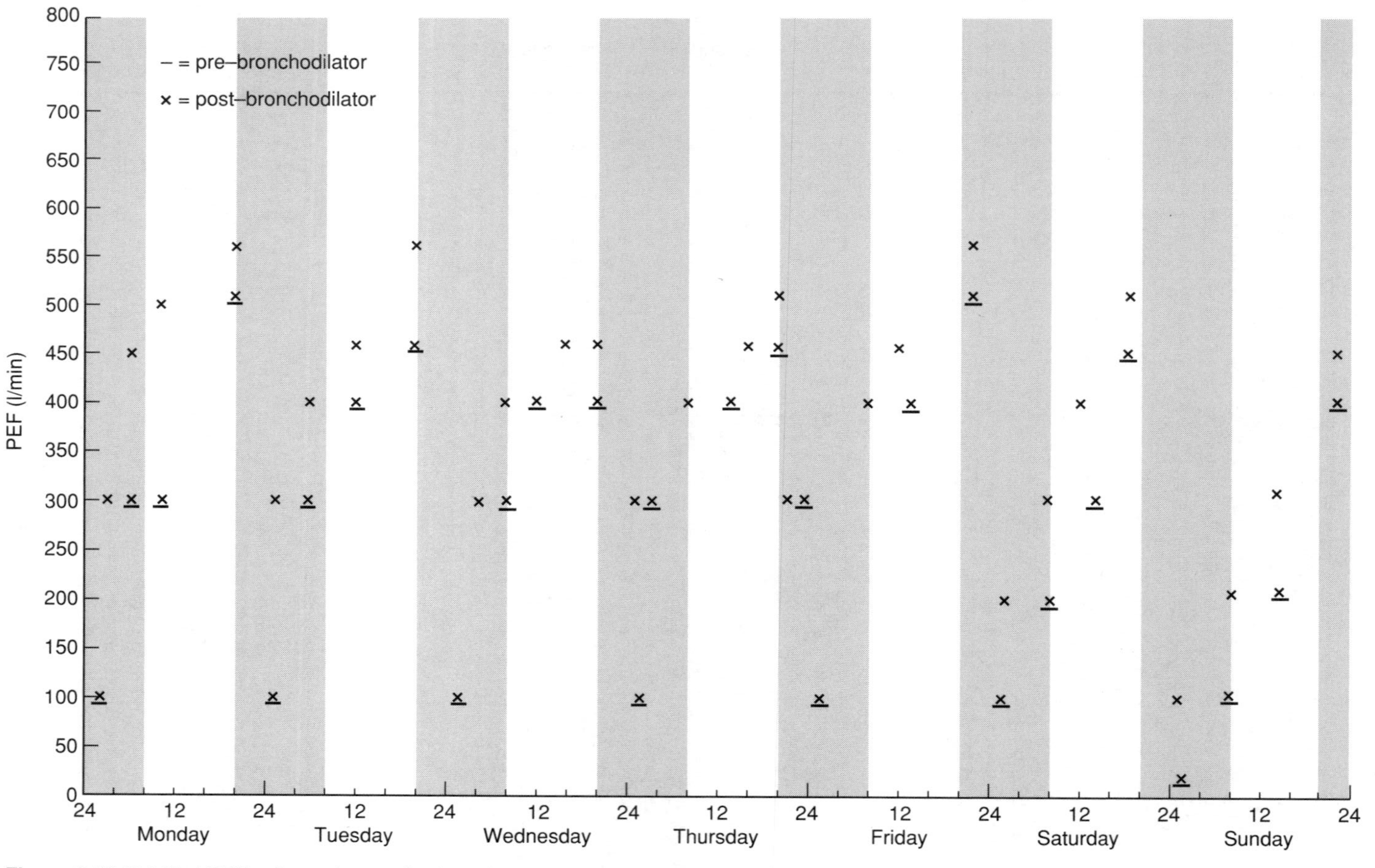

Figure 2.12 Falsified PEF values when patient's asthma 'worsens'.

asthma, is that of exaggeration of falls in peak flow or deliberate lowering of peak flow values during work. This results in sudden falls from Monday morning onwards, values remaining consistently low throughout the week but improving suddenly on a Friday evening. When analysed over a period of time, these variations appear to be too good to be true, although a genuine effect may then be documented by proceeding to bronchial challenge.

Handling the patient who is thought to be falsifying readings will depend on what may be the patient's reasons for manipulating the data. It is clearly important not to lose any trust and rapport which has been established, but explanation that false data could lead to significant mismanagement of their asthma or inability to identify an occupational cause often helps.

2.8 Absolute value or per cent predicted?

Over the years there has been much debate on whether absolute values of a lung function parameter should be used when assessing changes or comparing age groups, or whether the per cent predicted for that age group should be used. More recently further fuel has been added to the fire by Miller and Pincock who have shown that, at least for spirometric variables, the per cent predicted calculations for these lung function parameters get less accurate with age. They have therefore suggested that standardized residuals should be used as an indicator of deviation from normality. However, this view has yet to become widely used even though logically and intellectually it would seem the right road to take.

The attraction of avoiding absolute values when comparing peak flow readings from one group of patients to another, or indeed from one individual to another, is that it overcomes the problem of age, height and sex. Nevertheless, if groups of patients are matched for these variables – in practice often a difficult thing to do – absolute values can then be used.

A problem can also arise when deciding whether to express differences as the change in absolute terms or as a percentage change. Table 2.3 shows that an increase in peak flow of 40 l/min from a baseline of 200 represents a 20% increase, whereas the same absolute change in

Table 2.3 Absolute values versus per cent predicted

Pre-bronchodilator	*Post-bronchodilator*	*Absolute change*	*Per cent change*
200 l/m	240 l/m	+40 l/m	+20%
400 l/m	440 l/m	+40 l/m	+10%

peak flow in someone with a baseline of 400 represents only a 10% improvement, which may not be regarded as a significant bronchodilator effect. On the other hand that 40 l/min may be crucially important in terms of symptoms for the patient with the lower baseline peak flow.

Consequently, it is important when considering clinical and research data that both absolute values and values in terms of percentage change are given so that other individuals reading and interpreting the data are able to draw their own conclusions.

2.9 Computerized analysis

There are a number of computer analyses available for analysis of peak flow readings in the research setting.

2.9.1 Cosinor analysis

The Cosinor method of determining diurnal variation in peak flow uses a least squares method to test the goodness of fit of the data to a sinusoidal waveform. A summary of the method can be found in Appendix D.

2.9.2 Regression analysis

A number of workers have produced regression equations which can be used to predict values of peak flow in any individual from a given population. These studies have mostly been made in asymptomatic non-smokers. The equations vary considerably and depend on the size and age distribution of the original normal population studied and their race. Consistent factors which are present in all the equations are age and height, bearing in mind that different equations are produced for the male and female populations. Occasionally weight is found to be of importance. It is, of course, appropriate to use the relevant regression equation for the population under study but, in European Caucasian populations, the ECCS tables are increasingly being used. The regression equations established by the ECCS are:

Adult male PEF = $6.146h - 0.043a + 0.154$
Adult female PEF = $5.501h - 0.030a - 1.106$
(h = height in metres; a = age in years)

There has been criticism of the ECCS regressions because the normal subjects studied included smokers and give consistently lower values than from a similar population of non-smokers. However, with increasing use of the ECCS values, comparisons between separate studies can be made.

A summary of other regression equations for different patient groups are given in Appendix E.

2.9.3 Cusum analysis

This form of analysis has been used in the past but has limited application.

The method involves calculating the cumulative sum of sequential changes in peak flow from an initial reference value, but is probably little better than visual analysis of the raw data.

It may prove of use:

1 where there is a wide scatter and small therapeutic gains are considered clinically important;
2 when looking for small gains, e.g. when adding a further drug to an existing multi-drug regime;
3 for early detection of improvement, deterioration or cessation of improvement during treatment.

Rigorous studies of this technique have not been performed.

References and further reading

Bellia, V., Cibella, F., Migliara, G., Peralfa, G. and Bonsignore, G. (1985) Characteristics and prognostic value of morning dipping of peak expiratory flow rate in stable asthmatic subjects. *Chest*, **88**, 89–93.

Burdon, J.G.W., Juniper, E.F., Killian, K.J., Hargreave, F.E. and Campbell, J.M. (1982) The perception of breathlessness in asthma. *Am. Rev. Respir. Dis.*, **126**, 825–8.

Cochrane, G.M. and Clark, T.J.H. (1975) A survey of asthma mortality in patients between 36 and 64 in the Greater London hospitals in 1971. *Thorax*, **30**, 300–5.

Connolly, C.K. (1987) Falsely high peak expiratory flow readings due to acceleration in the mouth. *Br. Med. J.*, **194**, 285.

Greenough, A. (1990) Are we recording peak flows properly in young children? *Eur. Respir. J.*, **31**, 1193–6.

Gregg, I. (1964) The measurement of peak expiratory flow rate and its application in General Practice. *J. Coll. Gen. Pract.*, 7, 199–214.

Henderson, A.J.W. and Carswell, F. (1989) Circadian rhythm of peak expiratory flow in asthmatic and normal children. *Thorax*, **44**, 410–14.

Hetzel, M.R., Clark, T.J.H. and Branthwaite, M.A. (1977) Asthma: analysis of sudden death and ventilatory arrests in hospital. *Br. Med. J.*, **i**, 808–11.

Hetzel, M.R., Williams, I.P. and Shakespeare, R.M. (1979) Can patients keep their own peak flow records reliably? *Lancet*, **i**, 597–9.

Miller, M.R. and Pincock, A.C. (1988) Predicted values: how should we use them? *Thorax*, **43**, 265–7.

Mitchell, D.M., Collins, J.V. and Morley, J. (1980) An evaluation of Cusum analysis in asthma. *Br. J. Dis. Chest*, **74**, 169–74.

Ormerod, L.P. and Stableforth, D.E. (1980) Asthma mortality in Birmingham 1975–77: deaths. *Br. Med. J.*, **280**, 687–90.

Prior, J.G. and Cochrane, G.M. (1980) Home monitoring of peak expiratory flow rate using Mini-Wright peak flow meter in diagnosis of asthma. *J. R. Soc. Med.*, **73**, 731–3.

Rubinfield, A.R. and Pain, M.C.F. (1976) Perception of asthma. *Lancet*, **i**, 882–4.

Shim, C.S. and Williams, H.M. (1980) Evaluation of the severity of asthma: patients versus physicians. *Am. J. Med.*, **68**, 11–13.

Turner-Warwick, M. (1977) On observing patterns of airflow obstruction in chronic asthma. *Br. J. Dis. Chest*, **71**, 73–86.

Patterns of peak flow

3

'Kitty has no discretion in her coughs,' said her father: 'She times them ill.'
'I do not cough for my own amusement,' replied Kitty fretfully.

Jane Austen, *Pride and Prejudice*, ch. 2

In this chapter we describe some of the commoner patterns of peak flow to be found in the diagnosis and management of patients with asthma.

3.1 Diagnosis

Asthma may present as episodic or persistent breathlessness, with or without wheeze, chest tightness, cough (either dry or productive), or as any combination of these symptoms. Consequently, repeated use of the peak flow meter on a domiciliary basis is an essential part of the investigation of patients presenting with new respiratory symptoms. Single measures of peak flow can be misleading (section 2.2.1).

3.1.1 Unexplained breathlessness (Fig. 3.1)

> An 11-year-old boy presented with a history of recurrent breathlessness without wheeze, episodes occurring on most days. His grandmother would not believe he had asthma because he did not wheeze. Peak flow charts showed a range of readings from 190 to 340 over a 2-week period, thus confirming the diagnosis of asthma. Introduction of appropriate anti-asthma inhaled therapy resulted in control of his symptoms and peak flow variability.

Equally, level peak flows over a 4-week period while symptoms persist should point one away from a diagnosis of asthma. This can be

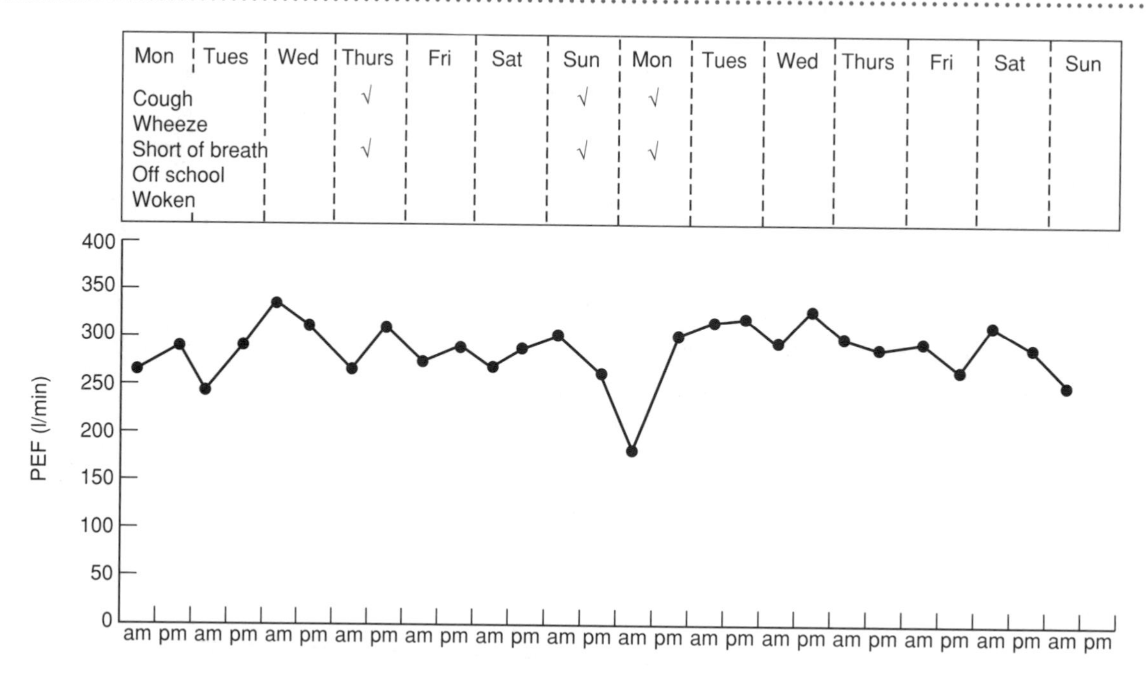

Figure 3.1 Peak flow record of an 11-year-old boy with episodic attacks of breathlessness.

helpful in the management of the chest tightness and hyperventilation commonly found in patients subject to panic attacks.

3.1.2 Cough (Fig. 3.2)

> A 40-year-old woman was referred to the Chest Clinic by her general practitioner with a 2-year history of paroxysmal coughing. She had a past history of tuberculosis but denied breathlessness. The cough only occasionally woke her from sleep but it was usually, though not invariably, worse on waking in the morning.

Peak flow charts showed a maintained value of around 220 l/min during the day, but with falls to around 150 l/min in the morning. The patient had been provided with a salbutamol inhaler and on mornings where peak flow was recorded after use of inhaled salbutamol this 'morning dip' was not as marked. Use of inhaled salbutamol on a relief basis did not control her symptoms (which is often the case where cough is the predominant symptom) but they improved with subsequent use of inhaled steroids.

3.1.3 Episodic symptoms

Occasionally patients present with symptoms that are infrequent, maybe occurring only once every 2–4 weeks, but cause considerable

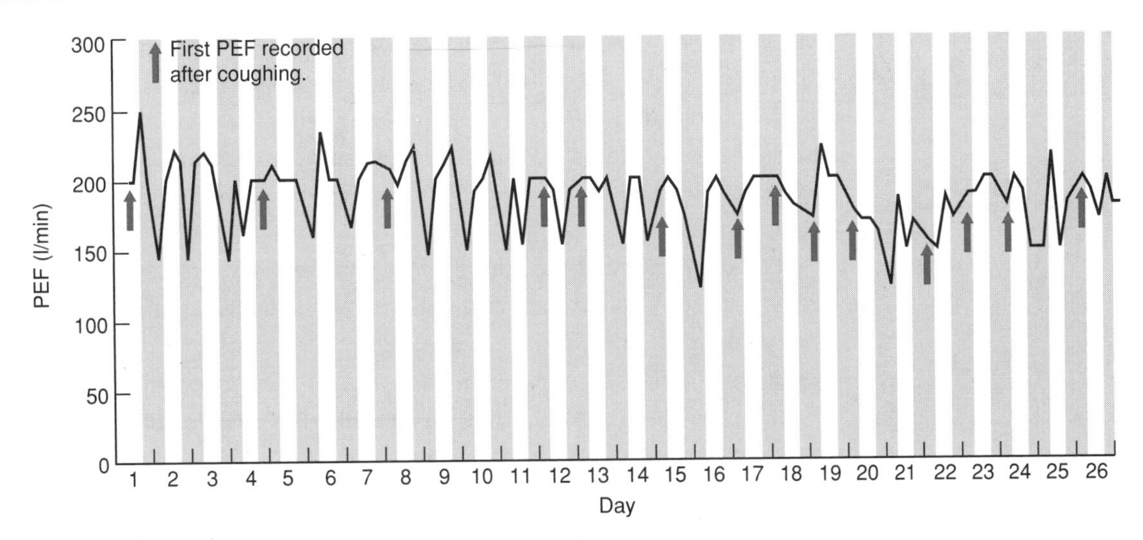

Figure 3.2 Peak flow record of a woman presenting with episodic cough.

symptomatic problems when they do occur. This history often turns out to be asthma and the peak flow meter is invaluable in making this diagnosis but it is important to identify that falls in peak flow coincide with the patient's symptoms.

3.1.4 Asthma versus chronic airflow obstruction

Variation in peak flow occurs both in patients with asthma and in patients with chronic airflow obstruction. There is a much discussed overlap between asthma and chronic obstructive bronchitis as clinical entities but when assessing an individual with airflow obstruction and a history of variable wheeze or breathlessness, domiciliary peak flow monitoring is often performed in an attempt to identify an asthmatic component. There are a number of problems inherent in doing this.

First, such a patient will have a fixed degree of airflow obstruction and so their peak flow will be below predicted value. This can lead to a false sense of security if used alone as a one-off measurement. In patients with emphysema, peak flow may often be relatively well maintained as it is achieved so quickly after the onset of a forced expiration, before the typical flow-dependent airway collapse seen in emphysema reduces airflow sharply thus permitting only a small expired volume of air in the initial record of expiration (i.e. FEV_1). The need for spirometry in this situation is crucial (section 1.3.2).

Secondly, peak flow is more likely to show a consistent, relatively invariable pattern in patients with chronic airflow obstruction, but if variability is present, the pattern of variation will not enable an asthmatic to be definitively identified from the patient with chronic airflow obstruction. There is evidence that patients with chronic airflow obstruction are more likely to show a combined morning and evening dip pattern than patients with asthma but this is by no means diagnostic.

Thirdly, once treated, the patient's pattern of peak flow variability may change. The patient with apparently fixed airflow obstruction and flat peak flows may often turn out to be overtly asthmatic with variable peak flow and with significant increases in peak flow and FEV_1, reinforcing the importance of being aware of what treatment the patient is taking when interpreting peak flow records.

It is often important to know, particularly in the context of recruitment of patients into research or drug trial protocols, the degree of peak flow variability or of reversibility to a bronchodilator. In patients with asthma, peak flow variability or bronchodilator response correlates well with other indices (e.g. bronchial reactivity, symptoms or response to treatment) but the picture in patients with chronic airflow obstruction is not so clear. In asthma (section 2.4), peak flow variability is most sensitively expressed as amplitude % mean, for epidemiological studies, whereas in chronic airflow obstruction a wide variety of ways of expressing variation correlates to mean peak flow. However, in chronic airflow obstruction these indices do not correlate well with clinical indices of disease activity or severity.

Consequently, use of peak flow in assessing response to treatment in chronic airflow obstruction is much less helpful than in asthma. Indeed, improvement in symptoms and in vital capacity (an index of deflation consequent upon improved small airway flow) may be marked when peak flow, either as an absolute value or in terms of variability, remains steadfastly unchanged.

Peak flow monitoring should therefore be used with caution in patients with chronic airflow obstruction and should always be backed up with spirometric information.

3.1.5 'Cardiac asthma'

In his monograph of 1835 on cardiac disease, James Hope noted that 'an immense proportion of asthmas . . . result from disease of the heart'. He continued that if this cause is not recognized the patient 'is harassed with a farrago of inappropriate remedies'. These words are still applicable today.

Wheeze can be caused by bronchial wall oedema resulting from the pulmonary venous hypertension of heart failure. This can cause a reduction in peak flow leading to less appropriate treatment for bronchial asthma.

> A 67-year-old man presented with breathlessness, wheeze and with a diurnal variation in peak flow sufficient to diagnose asthma (Fig. 3.3). He showed some response to inhaled β2-agonist but failed to improve on full asthma therapy. He was admitted to hospital following a blackout, where complete heart block was diagnosed and a pacemaker inserted. His peak flow rose to normal levels with minimal variability indicating that the cause of his low and variable peak flow was heart failure. He had presumably been switching in and out of heart block in the long term.

Care must be taken in the interpretation of peak flow records in patients who may have overt or occult cardiac disease. If an older patient with significant diurnal variation in peak flow fails to respond to standard inhaler therapy, a trial of diuretics would be worth while.

3.2 Events

When a patient is recording peak flow at home, variation in flow either gentle or abrupt, will occur for a variety of reasons. In some cases, the cause will be obvious (e.g. viral infection, exposure to an allergen) but in others less so. This variability may reflect inadequate underlying control and the physician should be aware of its relevance. Consequently it is always important to encourage the patient to record any possible trigger on the peak flow chart.

3.2.1 Seasonal changes

In general practice, most exacerbations of asthma are seen in the summer months coinciding with the pollen season, whereas hospital admissions for asthma are highest in the last quarter of the year, probably largely related to viral respiratory tract infections. In some patients, therefore, peak flow readings will decline during either season, and the patient can respond to, or pre-empt these declines by increasing their prophylactic treatment according to a management plan (Chapter 5).

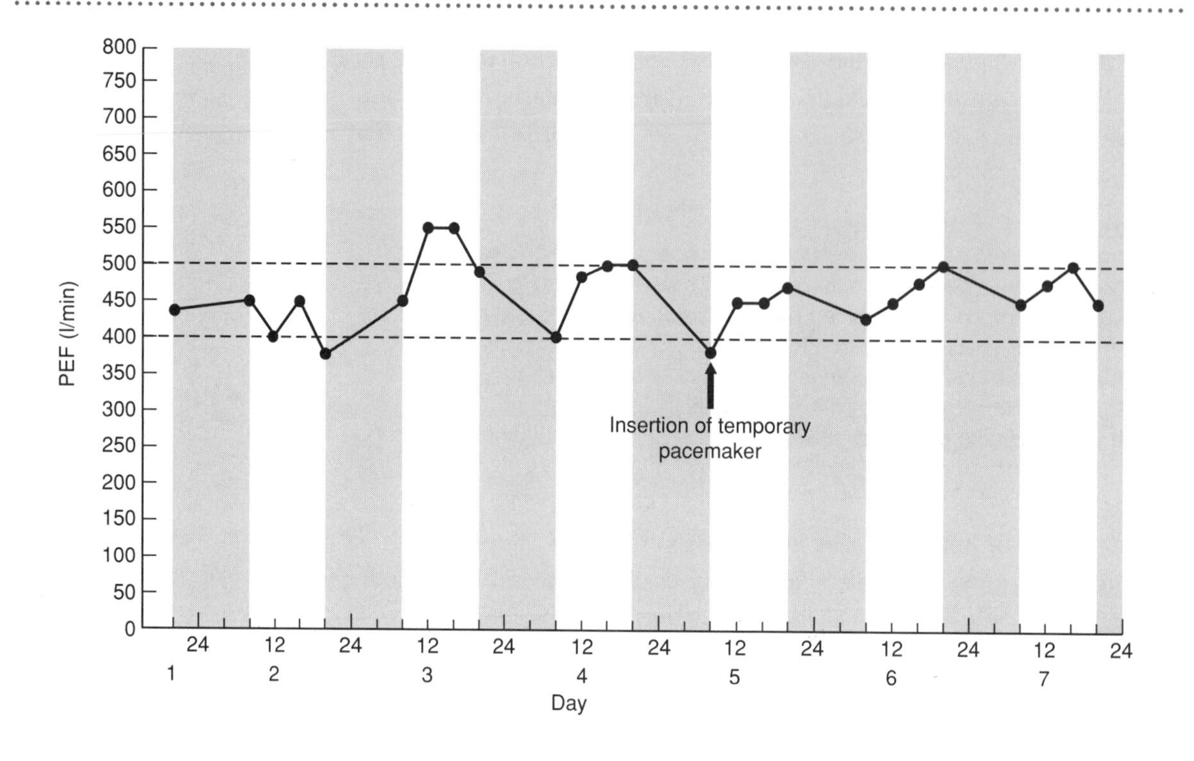

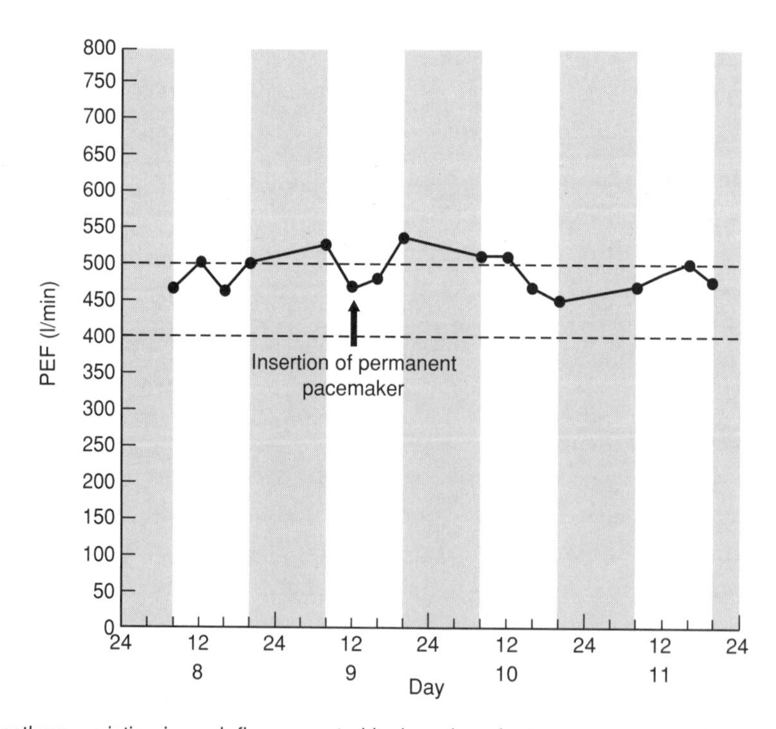

Figure 3.3 Cardiac asthma: variation in peak flow corrected by insertion of a temporary pacemaker.

3.2.2 Allergen exposure

Chronic exposure of a sensitized individual to an allergen such as the house dust mite will be reflected in their overall asthma control and may not necessarily show identifiable drops in peak flow from that patient's usual norm. However, exposure to a much higher level than usual, such as entering a particularly dusty room or when vacuuming or dusting, may result in a fall in peak flow which may be immediate but often occurs more over the subsequent few hours and may even be associated with a lower peak flow on the following morning.

Exposure to aero-allergens which are met only intermittently, may have a much more dramatic effect (Fig. 3.4).

An 18-year-old girl, known to be allergic to dogs, entered a friend's house for half an hour. Although the friend had removed the dog from the room the patient developed runny nose and eyes and had to leave after 30 minutes by which time she had become wheezy, peak flow having fallen to 290 l/min from her normal 500 l/min. Recovery of peak flow had not occurred by the following evening, suggesting that this was an immediate (Type I) allergic response with a significant delayed (Type III) effect.

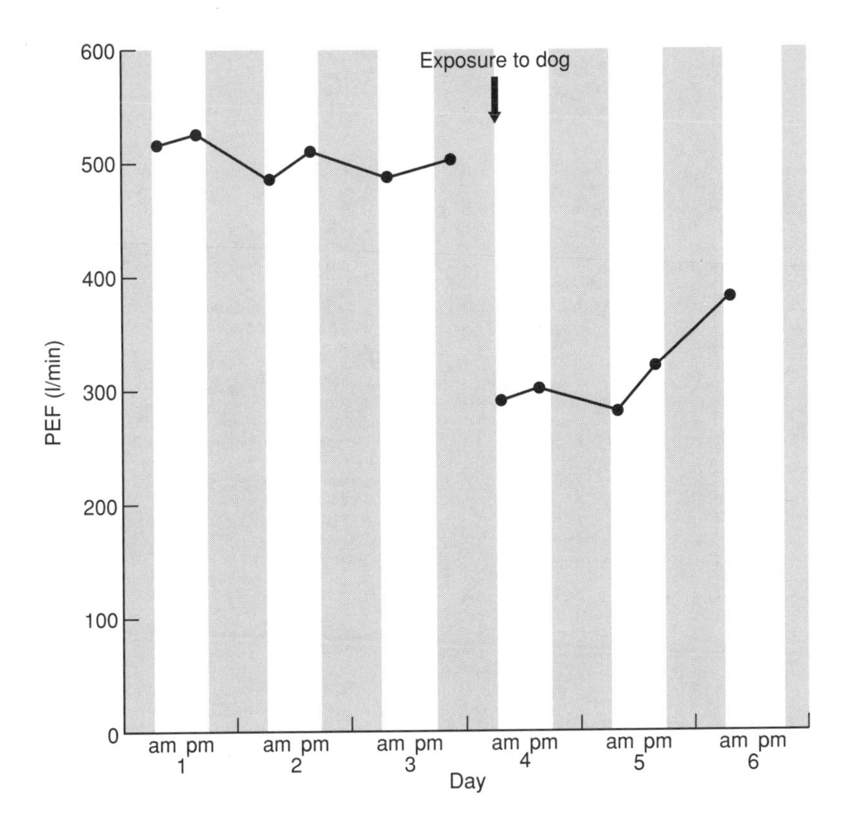

Figure 3.4 Forty per cent fall in peak flow after dog exposure in a sensitized individual.

Ingested allergens can also cause acute falls in peak flow although the prevalence of true food allergy amongst the asthmatic population is undetermined. A wide range of foods and both alcoholic and non-alcoholic drinks have been shown to cause asthma attacks in susceptible individuals.

> A 20-year-old girl with brittle asthma since childhood was admitted to hospital for stabilization of her asthma. Early one evening she developed acute worsening of her asthma with a fall in peak flow. On persistent questioning she admitted she had eaten two chocolate bars that afternoon knowing that chocolate invariably made her more wheezy (Fig. 3.5).

3.2.3 Viral infections

Upper respiratory tract infections are more likely to result in lower respiratory tract symptoms in patients with asthma. Even when symptoms are limited to nose and throat, however, peak flow can fall to a slight degree. Usually there is a lag before such a cold 'goes to the chest'

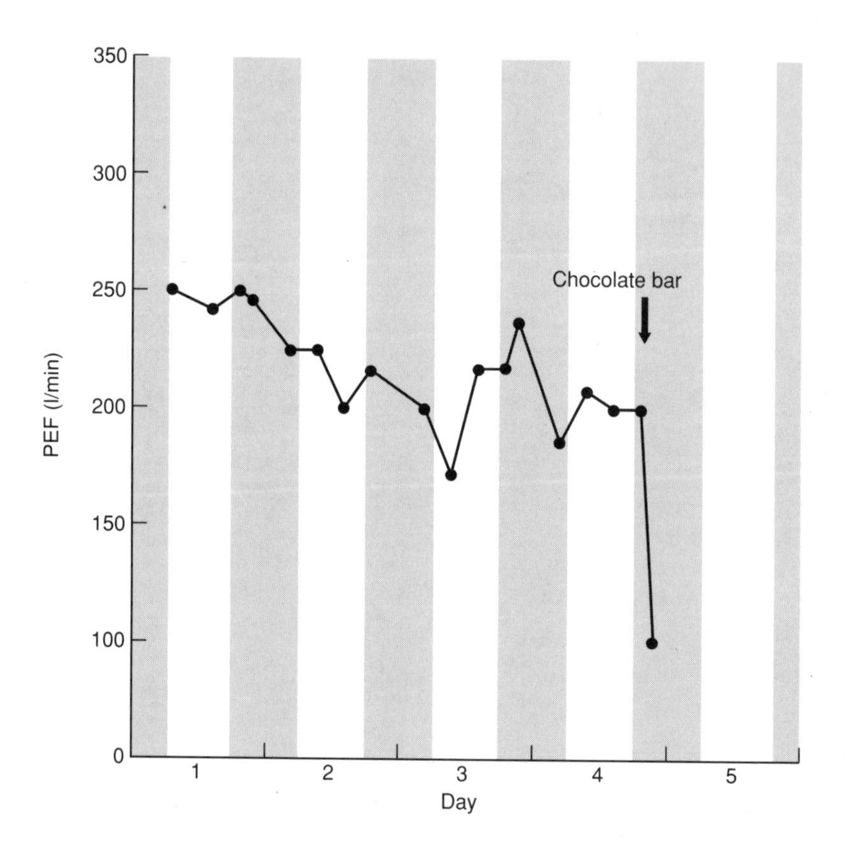

Figure 3.5 Asthma attack induced by eating chocolate. Note that peak flows had fallen slightly over the previous two days – patients with food allergy often report that food-induced attacks only occur when their asthma is generally worse than usual.

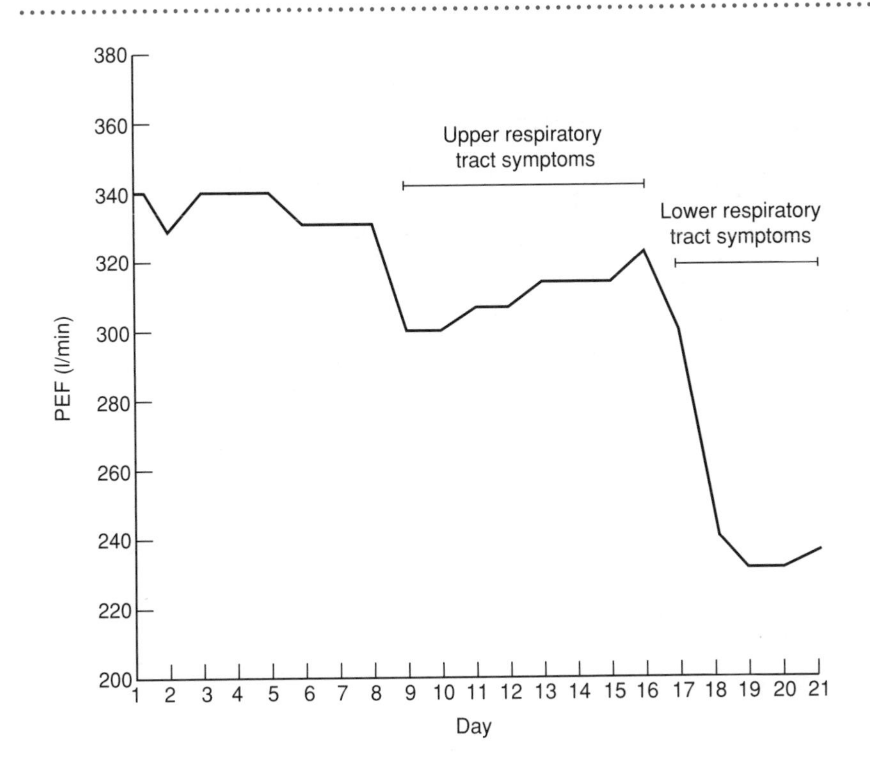

Figure 3.6 The double drop pattern of asthma worsened by a viral infection. An initial transient fall occurs during the phase of nasal symptoms followed a few days later by the infection 'going to the chest'.

and produces cough and/or wheeze (Fig. 3.6). This is one of the commonest presentations of asthma in general practice.

3.2.4 Change of environment

Many patients report effects on their asthma with changes in their environment. This can either be with changes in their micro-environment, e.g. emerging from a hot, steamy kitchen into a cool hallway; entering a hot, smoky bar from cool, relatively clearer outside air, or with a geographically more obvious change such as leaving their city home to holiday abroad. Both can result in changes in peak flow, whether transient or prolonged.

> A 35-year-old woman with longstanding but controlled asthma spent a Saturday lunchtime in a smoke-filled local pub. Peak flow on going to bed was lower than usual, she woke once that night needing relief β-agonist and had a more marked morning dip. She had recorded her 'smoke exposure' on her peak flow chart thus enabling identification of the cause of deterioration. Having usually run at 300 or above, her peak flow fell to 200. This is a highly significant fall from her normal state and such changes should not be brushed off as insignificant (Fig. 3.7).

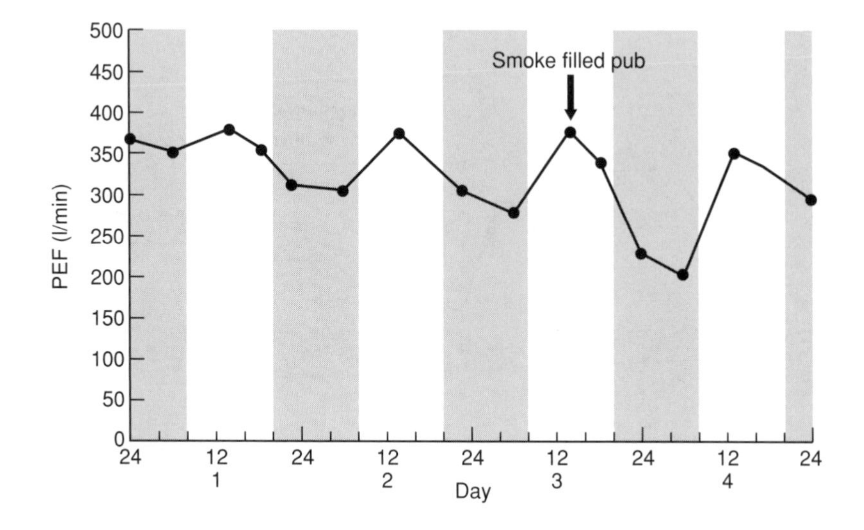

Figure 3.7 Delayed fall in peak flow with increased diurnal variation following exposure to a smoky environment.

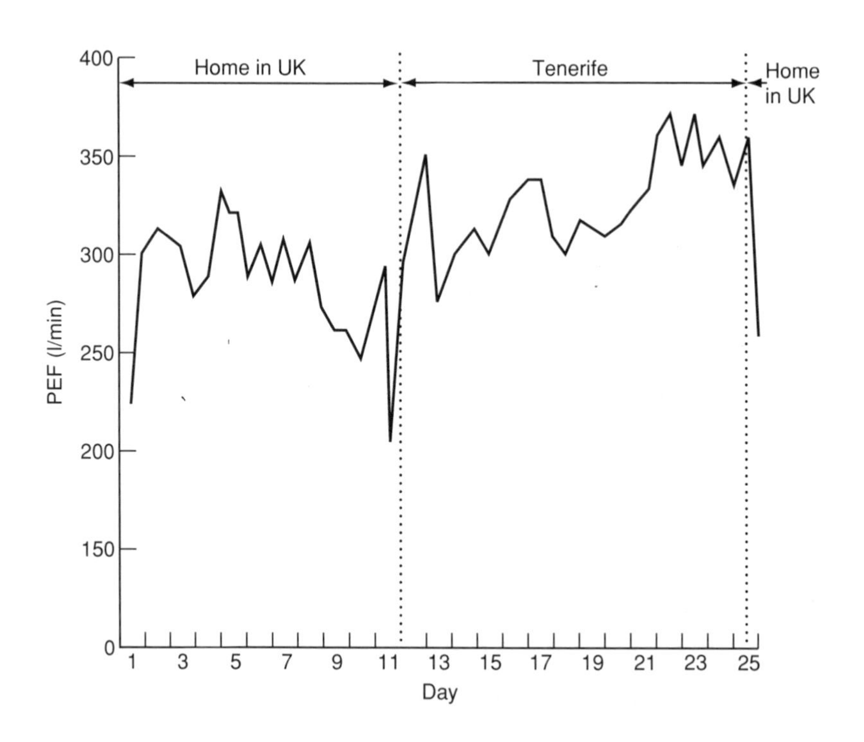

Figure 3.8 Improvement in peak flow on holiday.

A 38-year-old woman with severe asthma since childhood spends one month a year in Tenerife. Her peak flows improve from 300 to 350 l/min plus and her inhaled β-agonist use reduces to zero. Symptoms and peak flows return to their usual level within 36 hours of returning to her urban home (Fig. 3.8).

3.2.5 Exercise

Exercise is a common trigger for acute episodes of wheezing and breathlessness, particularly in children. On a day-to-day basis peak flows will not be recorded following exercise as use of relief bronchodilator rapidly restores the status quo in most cases. Exercise is used, however, to demonstrate bronchial lability diagnostically, in epidemiological research or clinical trials.

The classic pattern is for peak flow to rise immediately after exercise falling to a trough 10–15 minutes later. If a parent needs convincing that his/her child's exercise-induced breathlessness is due to asthma, a brisk run with peak flow readings before and at 2 to 5 minute intervals after exercise for 20 minutes can produce convincing evidence for the parent.

> An 11-year-old boy gave a history of cough and wheeze on exertion (i.e. games) for a year. He was exercised by free running for 4 minutes (Fig. 3.9).

3.2.6 Premenstrual asthma

This is quite common although the exact mechanism is far from clear. Occasionally the exacerbation may be very severe, sufficient to need hospitalization. In some women asthma deteriorates during the two days or so before menstruation while in others falls in peak flow are only

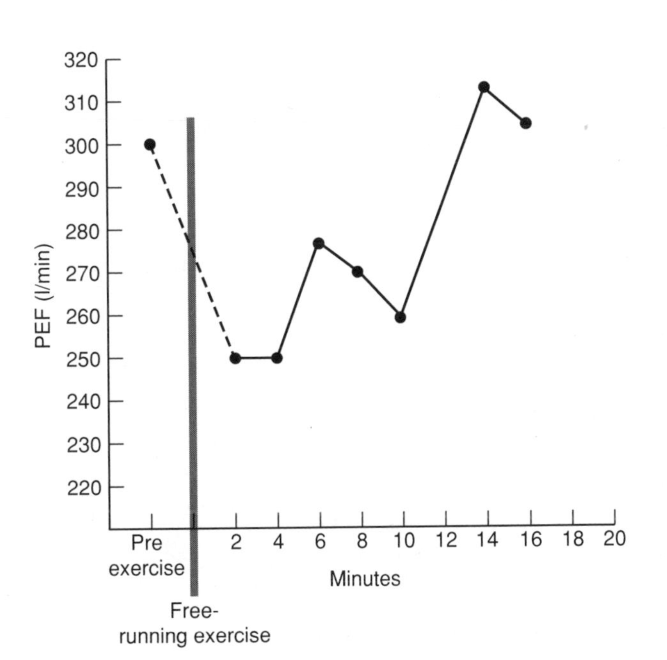

Figure 3.9 Exercise-induced asthma.

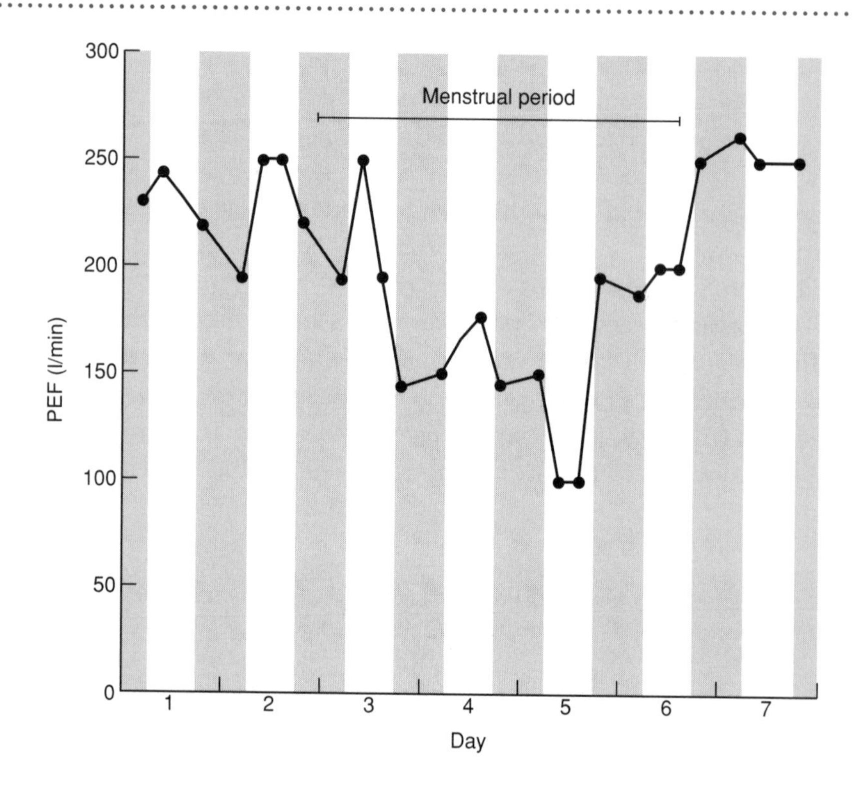

Figure 3.10 Asthma associated with the menstrual period.

seen after menstrual loss has begun. It is important that a patient records on what day menstruation occurs when looking for contributory factors to poor asthma control.

> A 46-year-old woman reported that her asthma usually, but not always, worsened at the beginning of her menstrual period (Fig. 3.10).

3.2.7 Drugs

Certain drugs are well known to cause asthma, such as aspirin and β-blockers (both selective and non-selective), either orally or as eye drops (Timolol). As β-blockers are usually taken as a regular medication they will act as a consistent contributor to asthma rather than as a cause of intermittent episodes. Conversely, aspirin and nearly all other non-steroidal anti-inflammatory agents, are often taken on an as-required basis and can thus cause acute severe episodes of asthma. If a peak flow record shows sudden unexplained falls, always ask for a history of analgesic consumption. Around 5% of patients with asthma in the UK are aspirin-sensitive and recognition of this potentially dangerous trigger is essential, although it is our advice to all asthmatics to avoid aspirin and other non-steroidal anti-inflammatory drugs if possible and use paracetamol for analgesia.

3.3 Treatment

Once the diurnal pattern of peak flow has been established in a patient with asthma, it is easy to observe, and act on, the effect of therapy on this pattern. In some cases, a low peak flow with little variability will respond by showing an increase in daily maximum value, the patient passing through a period of marked diurnal variation before further changes in treatment result in settling of the variation with a higher overall daily peak flow.

The way in which peak flow patterns improve and the rate at which they improve will depend on:

1 the patient;
2 the degree of airflow obstruction;
3 the type of treatment given.

3.3.1 Response to inhaled bronchodilators

The British Thoracic Society recommendations are that in mild asthma the initial treatment step is to use inhaled β-agonist on an 'as-required' basis. In these circumstances patients should be asked to record their peak flow before and after use of their inhaler (using different colours, see section 2.4) to assess the degree of improvement. However, one must beware of the patient who, when well, records peak flows when little relief inhaler is used but whose peak flows apparently increase when not so well, as peak flows are recorded after use of relief bronchodilator.

The degree of bronchodilatation is dose-dependent up to a certain level (which differs between patients) so a lack of bronchodilator response may only reflect that the dose used on that occasion was inadequate. Post-bronchodilator peak flows should be measured 10–15 minutes after inhaled β_2-agonist but 30–45 minutes after inhaled ipratropium.

3.3.2 Response to inhaled steroids

If on-demand bronchodilators are ineffective in controlling symptoms, or use exceeds one to two puffs (100–200 μg salbutamol) per day, an

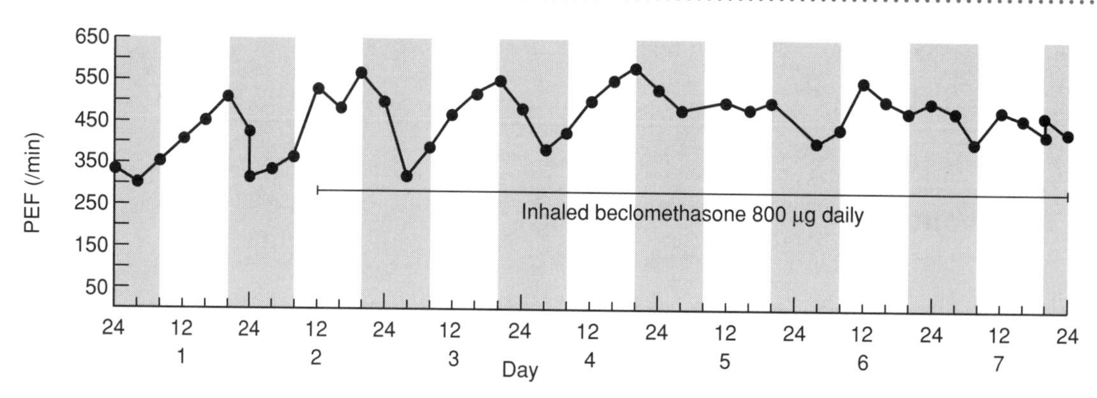

Figure 3.11 Effect of inhaled steroids on peak flow variability.

inhaled prophylactic agent is the next step (Fig. 3.11). It is useful to record daily use of β-agonist on the peak flow charts during introduction of a preventative treatment. It acts as a measure of improvement (as relief inhaler use falls) or can warn that a further increase in prophylactic agent is necessary, if there is continuing use of the relief bronchodilator.

3.3.3 Response to other prophylactic agents

The same general rules apply to the use of disodium cromoglycate or nedocromil sodium. Both are usually prescribed four times a day but either may confer a degree of control at thrice or twice a day; peak flow monitoring will enable this to be best assessed.

> A 65-year-old man had a long history of intermittent paroxysms of dry cough. Peak flow monitoring shows an initial diurnal variation of 13% with a reduction in diurnal variation and increase in mean peak flow on disodium cromoglycate 10 mg q.d.s. by metered dose inhaler (Fig. 3.12).

3.3.4 Response to oral theophyllines

Some patients appear to benefit from oral bronchodilators, such as the theophylline family or oral β-agonists. In many cases increasing inhaled prophylaxis allows removal of an oral bronchodilator from a patient's prescription but occasionally this is not possible. Peak flow monitoring is useful in reassuring both patient and doctor or nurse that removal or retention of oral theophylline is the best option.

> A 76-year-old man with late onset asthma displayed adequate inhaler technique using a large-volume spacer, but diurnal variation in peak flow only improved with oral slow release aminophylline (Fig. 3.13).

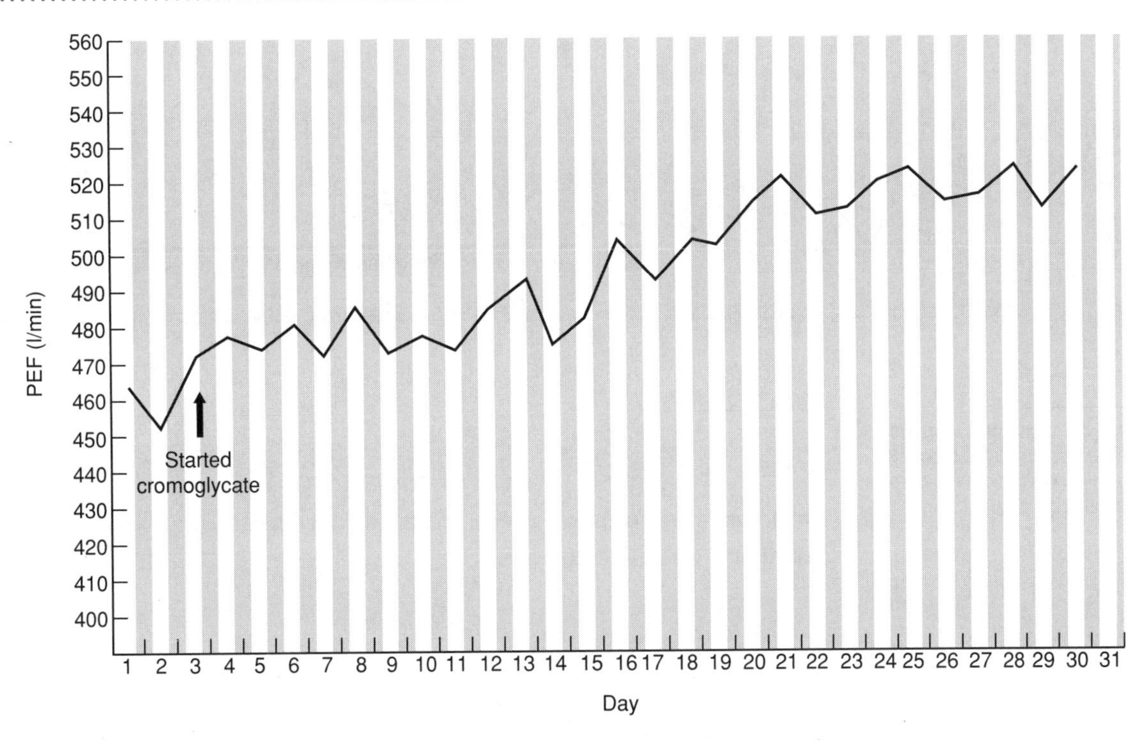

Figure 3.12 Improvement in peak flow following introduction of inhaled disodium cromoglycate.

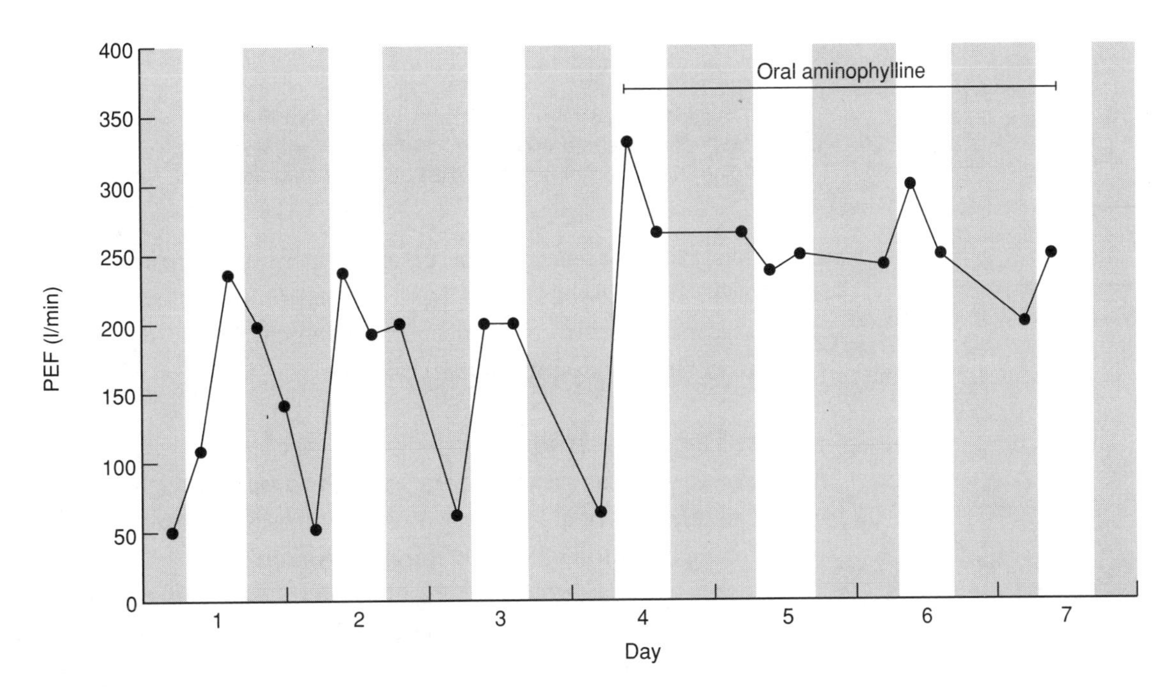

Figure 3.13 Improvement in morning dip after introduction of oral aminophylline.

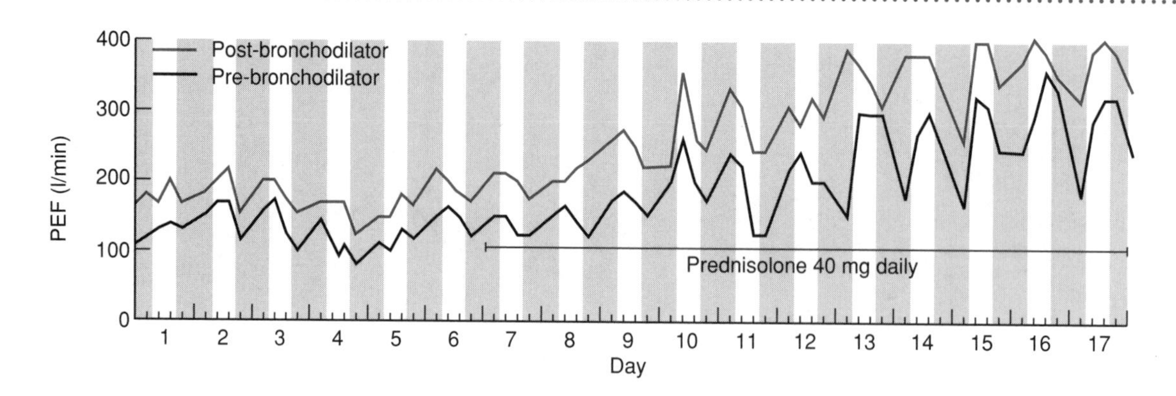

Figure 3.14 Gradual response to oral corticosteroids in a patient presenting with apparent chronic airflow obstruction.

3.3.5 Response to oral corticosteroids

Oral corticosteroids are used in three situations in the management of asthmatic airflow obstruction.

1. In acute asthma (see below).
2. As a steroid 'trial' in patients presenting with marked, long-standing airway narrowing.
3. To treat the patient whose symptoms and lung function have gradually declined over a period of months but who is not suffering an acute attack.

> A middle-aged woman with chronic, worsening breathlessness and low peak flows responded to high dose oral steroids over a period of two weeks, developing the classical morning dip pattern of asthma (Fig. 3.14).

Oral steroids may need to be given for periods of up to 4 weeks at doses of 40 mg daily or higher. In this situation peak flow monitoring is mandatory not only to monitor improvement but to detect further gradual worsening despite oral steroid use.

3.3.6 Use of nebulizer

Any patient who has asthma severe enough to warrant either intermittent or regular use of a nebulizer must monitor their peak flows before and after each nebulized dose of bronchodilator.

Patients should be told that with an acute attack, if the peak flow does not improve, they should inform their general practitioner urgently and administer a further nebulized dose, or call an ambulance or go to hospital direct.

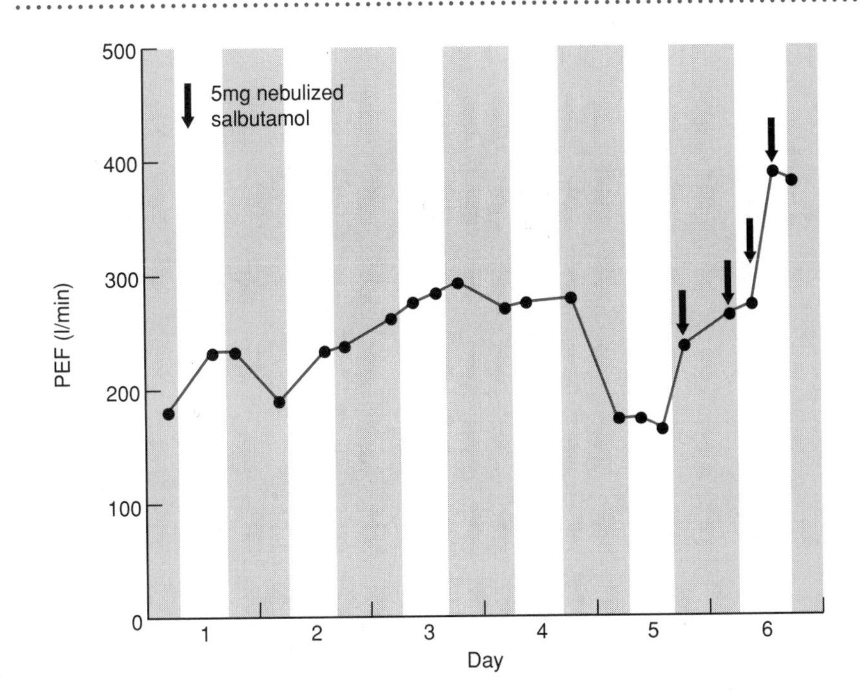

Figure 3.15 Improvement in peak flow in a patient with severe chronic asthma only by high dose repeated doses of nebulized salbutamol.

In some patients with severe asthma, successive nebulized doses of β-agonists can markedly improve peak flow (Fig. 3.15).

3.4 Acute asthma

The onset of acute asthma can be rapid (Fig. 3.16), occurring in a matter of minutes, or more gradual (Fig. 3.17), occurring over a day or two or more. Often the acute episode will have been predicted by the onset of nocturnal symptoms or increased use of β-agonists, the latter managing to keep peak flows at acceptable levels for a time before a severe attack finally occurs. The use of management plans (Chapter 5) will prevent many asthma attacks.

Following an attack there are different patterns of recovery. In some patients, particularly children, recovery may be very rapid, while in others, usually older patients or patients with longstanding severe asthma, the recovery may take place only over 2–3 weeks (Fig. 3.18). Most follow a middle course but often recover with the development of pronounced morning dipping.

Recovery from an acute attack can be either with (Fig. 3.19) or without (Fig. 3.20) morning dips.

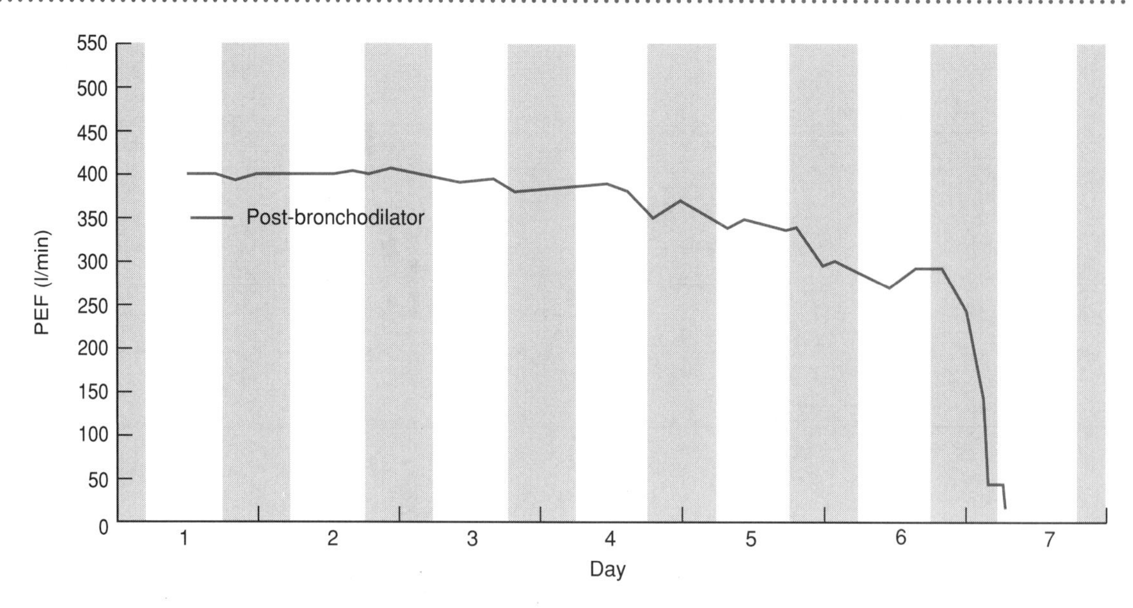

Figure 3.16 Rapid onset of an acute asthma attack.

The British Thoracic Society guidelines for the management of acute severe asthma state that discharge from hospital following an acute attack of asthma should not occur until diurnal variation in peak flow (max – min/max %) is less than 25%. The morning dip

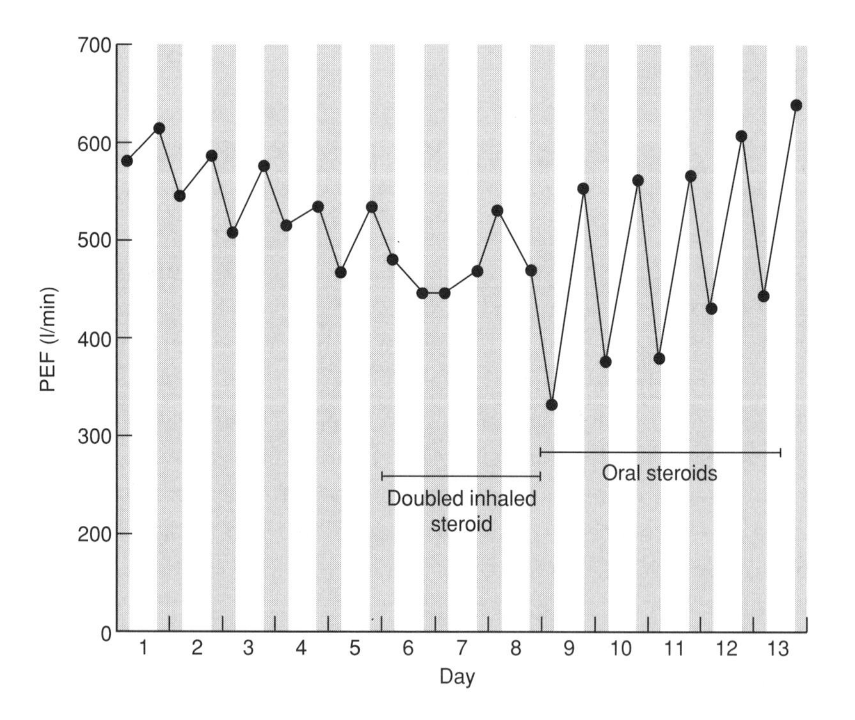

Figure 3.17 Gradual onset of an acute asthma attack.

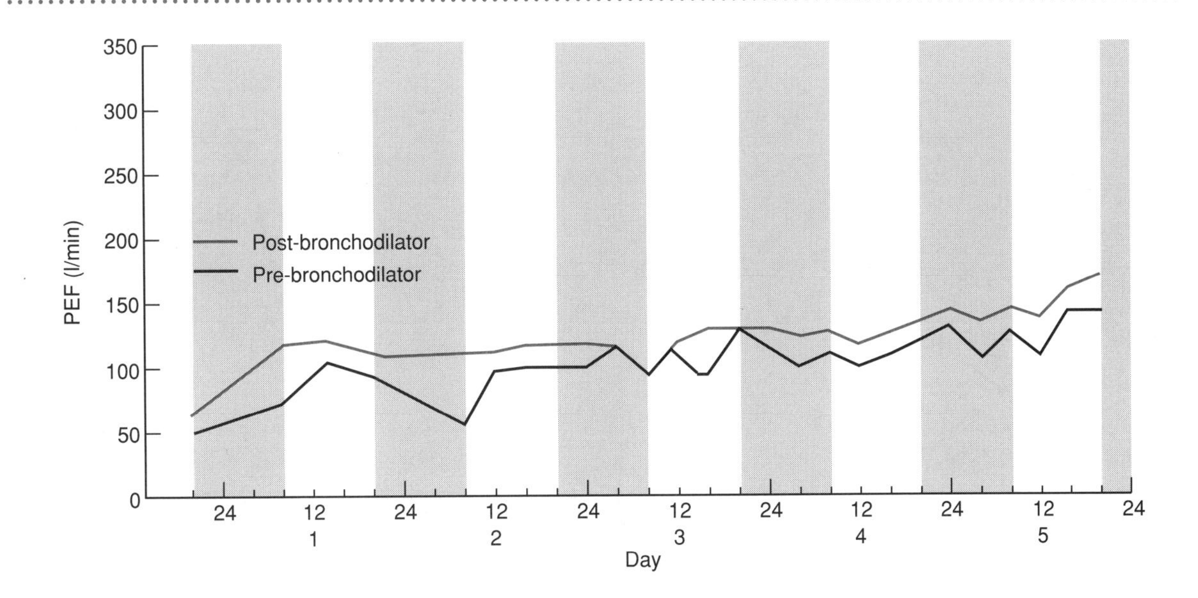

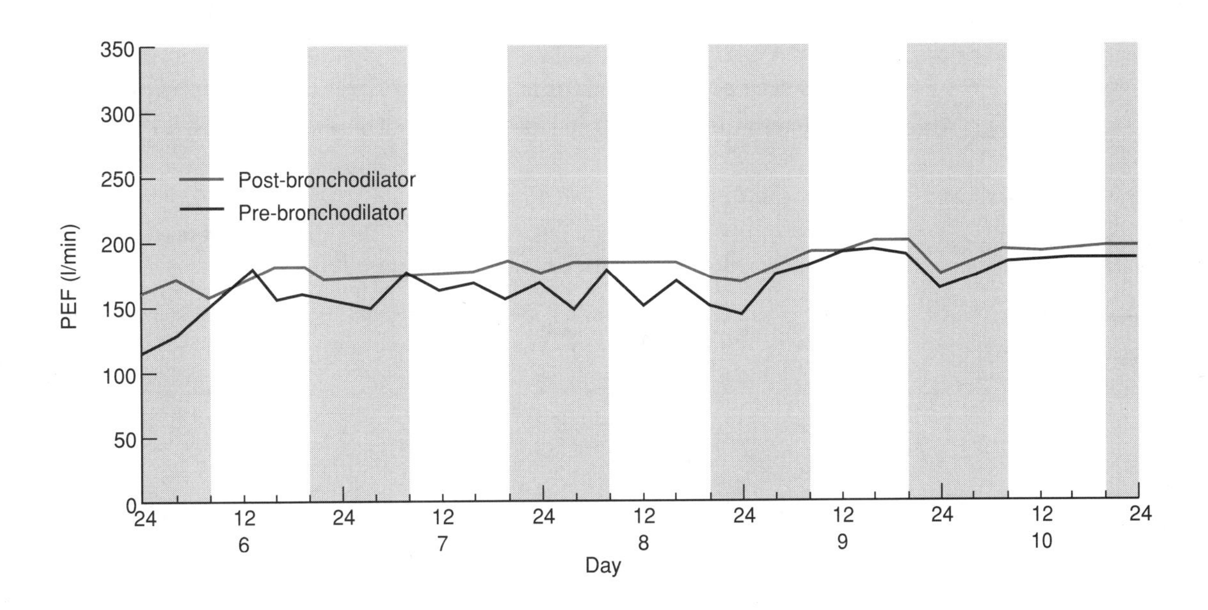

pattern seen during recovery from acute asthma is often a transient problem and in itself may not need treatment in its own right, although when it persists long-acting inhaled or oral bronchodilators can be used.

Figure 3.18 Gradual improvement in peak flow following an asthma attack in an elderly woman (predicted peak flow 280 l/min).

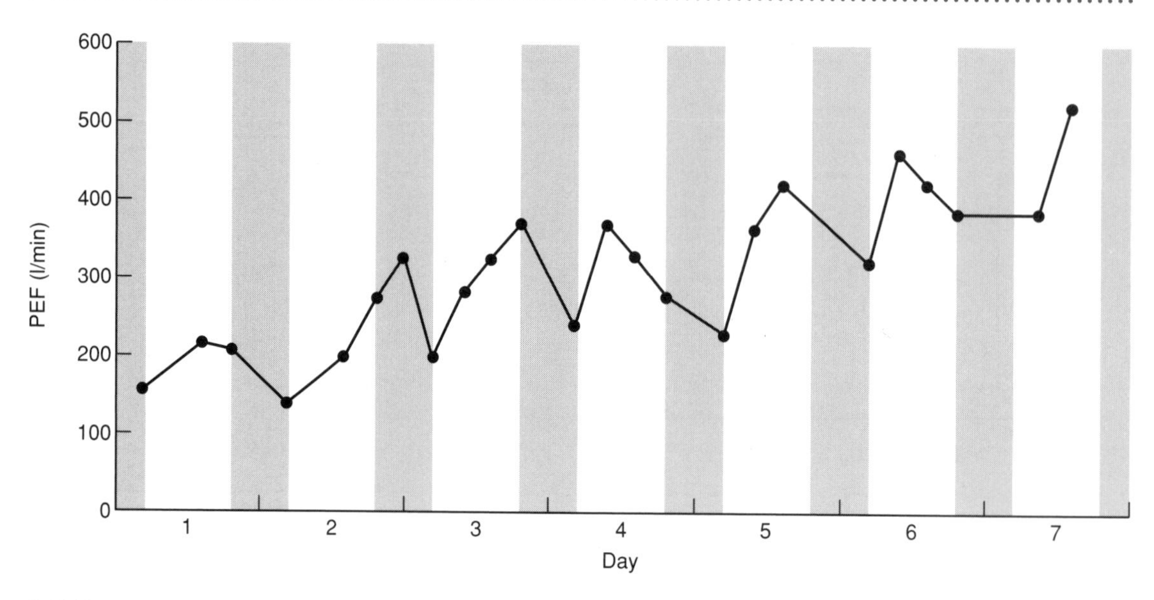

Figure 3.19 Recovery from acute asthma with pronounced morning dips. Diurnal variation of peak flow of less than 25% achieved by the sixth day.

3.5 Brittle asthma

This serious form of asthma is characterized by severe symptoms with recurrent attacks, often associated with wide swings in peak flow

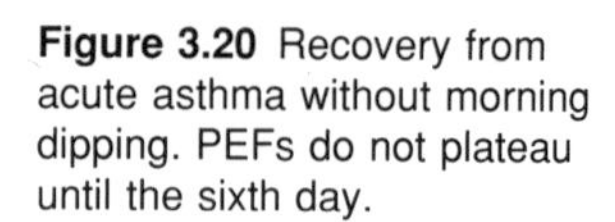

Figure 3.20 Recovery from acute asthma without morning dipping. PEFs do not plateau until the sixth day.

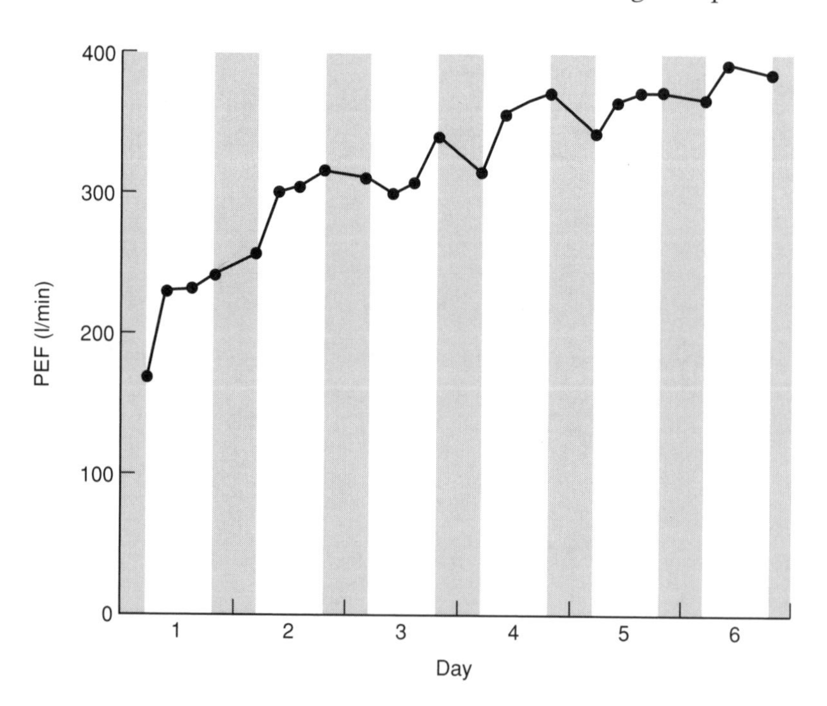

between attacks. The condition is difficult to treat although some may benefit from continuous infusions of subcutaneous β-agonists.

> In a 42-year-old woman with late onset, brittle asthma the pattern of peak expiratory flow is erratic (Fig. 3.21), with a diurnal variation exceeding 40% despite maximal medical therapy. Her peak flow variability and symptoms were controlled by continuous subcutaneous infusion of terbutaline (Fig. 3.22).

3.6 Traps

There are a number of traps into which the unwary may fall when interpreting peak flow data (see also Chapter 2). Although they do not occur very often, it is important to recognize them.

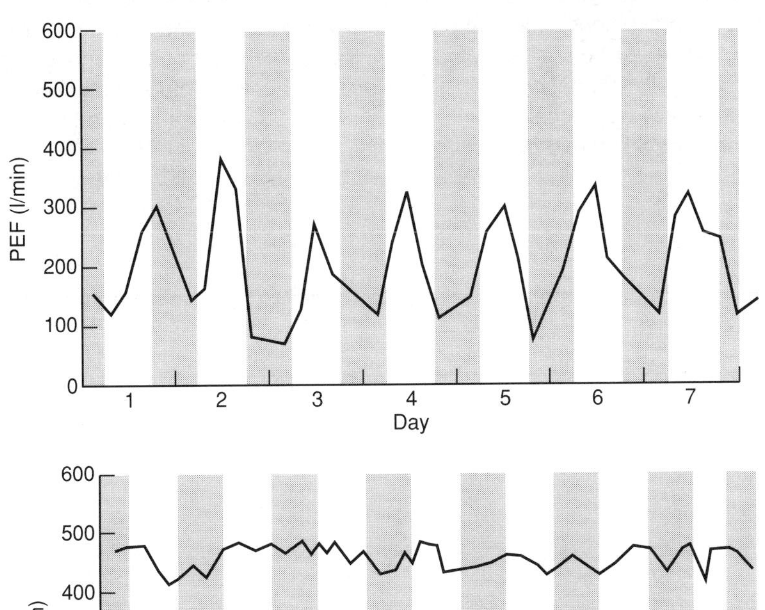

Figure 3.21 Brittle asthma showing chaotic swings in peak flow.

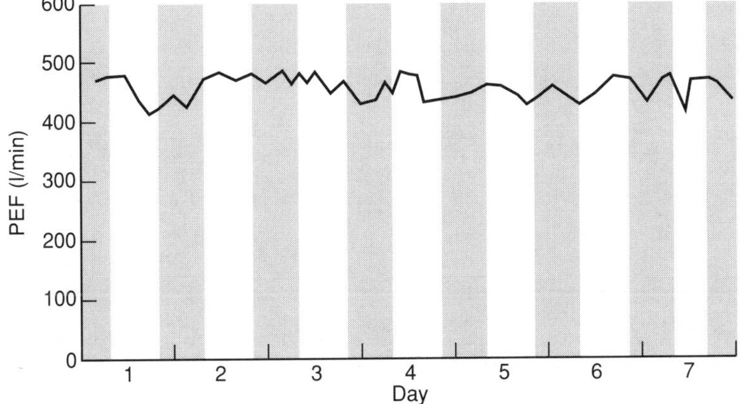

Figure 3.22 The same patient controlled by subcutaneous terbutaline.

3.6.1 Problems with the patient

The pain of pleurisy or a fractured rib will obviously inhibit someone attempting to record peak flow and measurement should be stopped until pain is relieved. In patients with a facial palsy air can sometimes leak around the mouthpiece, although they usually develop a 'sealing' mechanism to overcome this.

3.6.2 Problems with the meter

As described in Appendix A there may be a significant difference between any two given peak flow meters, either of the same or different makes. Lack of appreciation of this may lead to inappropriate adjustment in treatment.

> A 37-year-old woman with moderate asthma reported that her peak flow had fallen from her usual value of 400 to 330–350 l/min but without significant alteration in her symptoms. An increase in inhaled steroid dose produced no further improvement. Then the patient realized that she had changed her peak flow meter from a Mini Wright to a Vitalograph at around the time her peak flow fell. The patient recorded values for the two meters confirming that her peak flow had remained unchanged (Fig. 3.23), so she returned to her original inhaled steroid dose.

Figure 3.23 Differences in peak flow recorded by two different meters, the Mini Wright (MW) and Vitalograph (V). The daughter (aged 9) also had asthma, but the patient, being of a scientific bent, obtained comparable readings from her husband and son, neither of whom had asthma.

		Patient		Husband		Son		Daughter	
		MW	V	MW	V	MW	V	MW	V
Sunday	am	430	380	650	580	290	280	230	225
	pm	425	360	620	580				
Monday	am	430	380	650	650	330	290	280	250
	pm	440	390	635	585	320	295	295	250
Tuesday	am	435	380	650	650	335	285	280	230
	pm	440	380	630	580				
Wednesday	am	420	365	640	625	330	330	270	240
	pm	435	365	630	590	340	310	320	260
Thursday	am	430	370	620	620	330	320	310	260
	pm	430	380	620	570	340	290		
Friday	am	440	360			330	290	270	250
	pm	425	365	580	520				
Saturday	am	430	375	620	530	300	290	310	270
	pm	440	370	560	500				

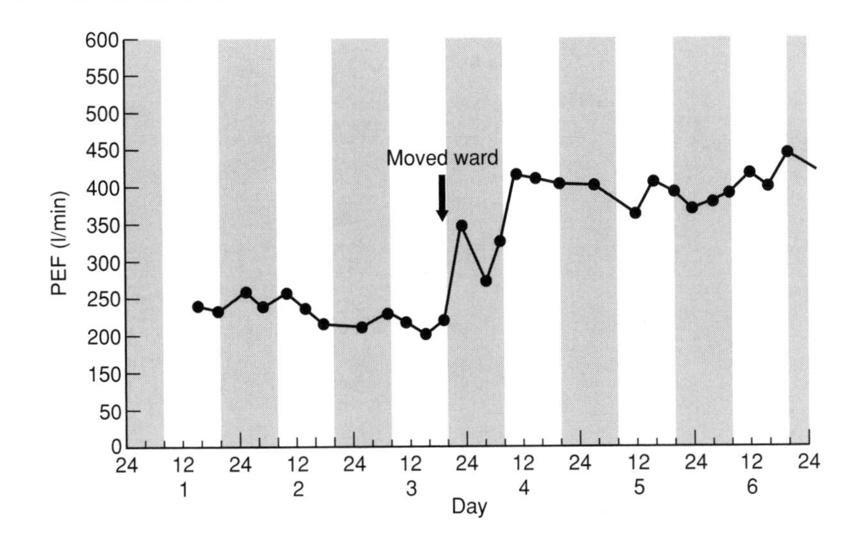

Figure 3.24 Faulty peak flow meter recording falsely low readings on the first ward where the patient was nursed. The problem was only identified when peak flow readings 'rose' when the patient was moved to a new ward and another peak flow meter used.

A 60-year-old man with asthma was admitted to a non-respiratory ward with acute asthma. After 3 days of unchanging peak flow readings the patient was transferred to a respiratory ward where peak flows apparently improved dramatically. This was shown to be due to an old, inaccurate peak flow meter on the non-respiratory ward (Fig. 3.24).

3.6.3 The 'good' peak flow

A 27-year-old weight-lifter was referred to the chest clinic with a history of episodic breathlessness on exertion. In the referring letter the GP wrote: 'He has a peak flow of 610 so this can't be asthma' (predicted peak flow was 560). Domiciliary peak flow monitoring showed variation between 610 and 710, an overall variation of only 14% (Fig. 3.25*a*). The history was good for asthma so he was started on inhaled steroids (beclomethasone 800 μg daily) with relief β_2-agonist as required. His symptoms resolved and his peak flow rose to between 700 and 800 (Fig. 3.25*b*).

This reinforces not only the danger of a single peak flow but also that 'predicted values' are guidelines and some individuals may, in a normal state, blow either significantly higher or lower than such a given value. It also reinforces that borderline diurnal variation should not be completely dismissed if the history is good for asthma.

3.6.4 Pseudo-loss of dipping

Airway narrowing occurring in the morning (morning dipping) may last only a brief time before resolving either with or without treatment.

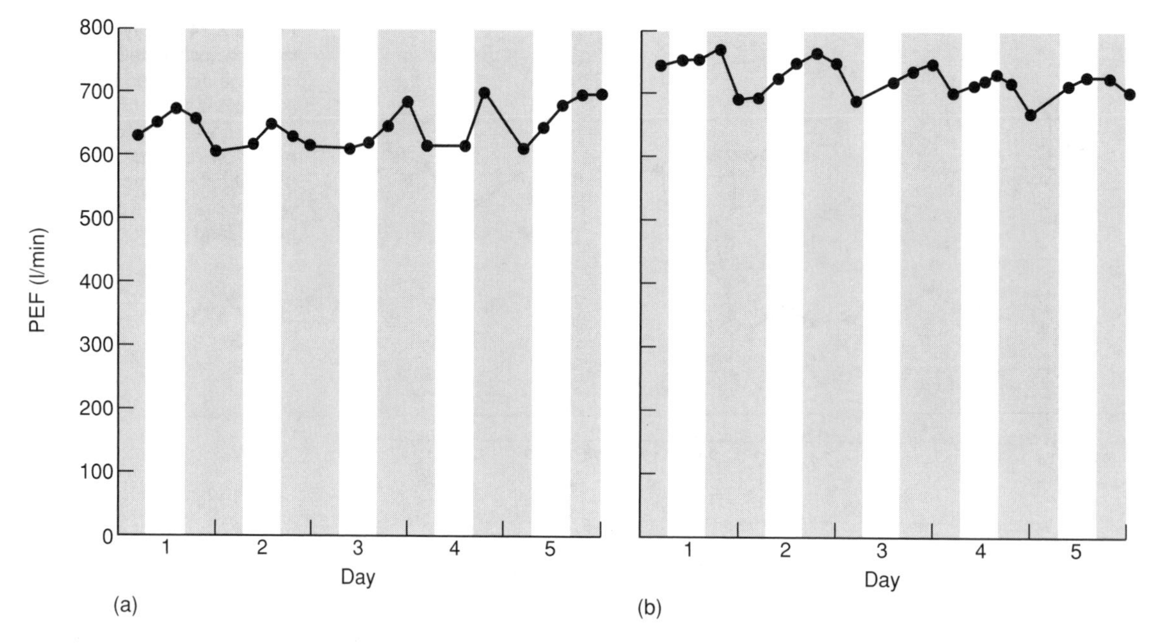

Figure 3.25 (*a*) Apparently 'good' peak flow readings (although always less than 700 l/min) in a weight-lifter which (*b*) improved by around 50 l/min (and always greater than 700 l/min) on inhaled corticosteroids with resolution of symptoms.

Consequently, changes in the time of performing the morning peak flow (e.g. later on a Sunday morning) may miss the morning dip. This was reported by Venables and her colleagues (1989) when investigating a case of suspected occupational asthma (Fig. 3.26).

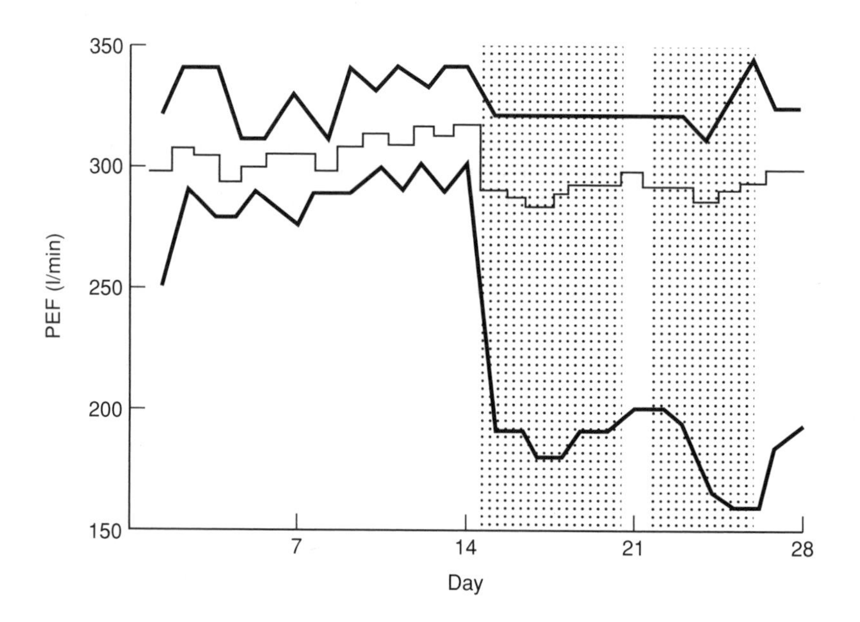

Figure 3.26 Daily maximum, mean and minimum peak flows recorded for two weeks while away from work followed by two separate weeks (*stippled areas*) at work. The lower daily minimum peak flow at work was due to the patient getting up much earlier on days at work (see Chapter 4, Occupational Asthma). (Reproduced with permission from Venables *et al.*, 1989.)

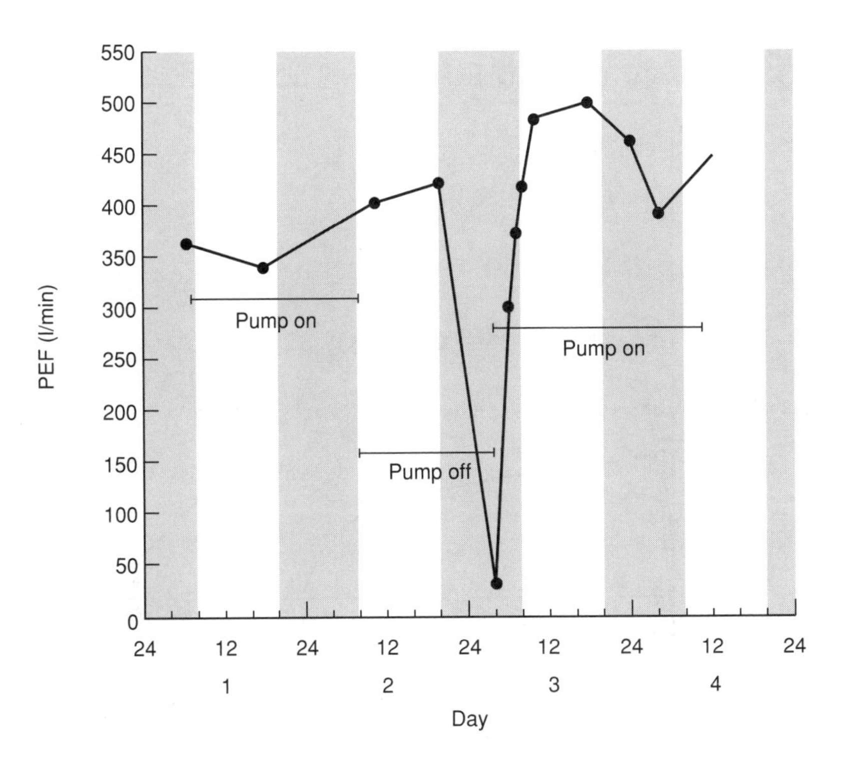

Figure 3.27 A patient with brittle asthma who disconnected his subcutaneous pump because of improving symptoms. Peak flow fell immediately and the pump had to be restarted.

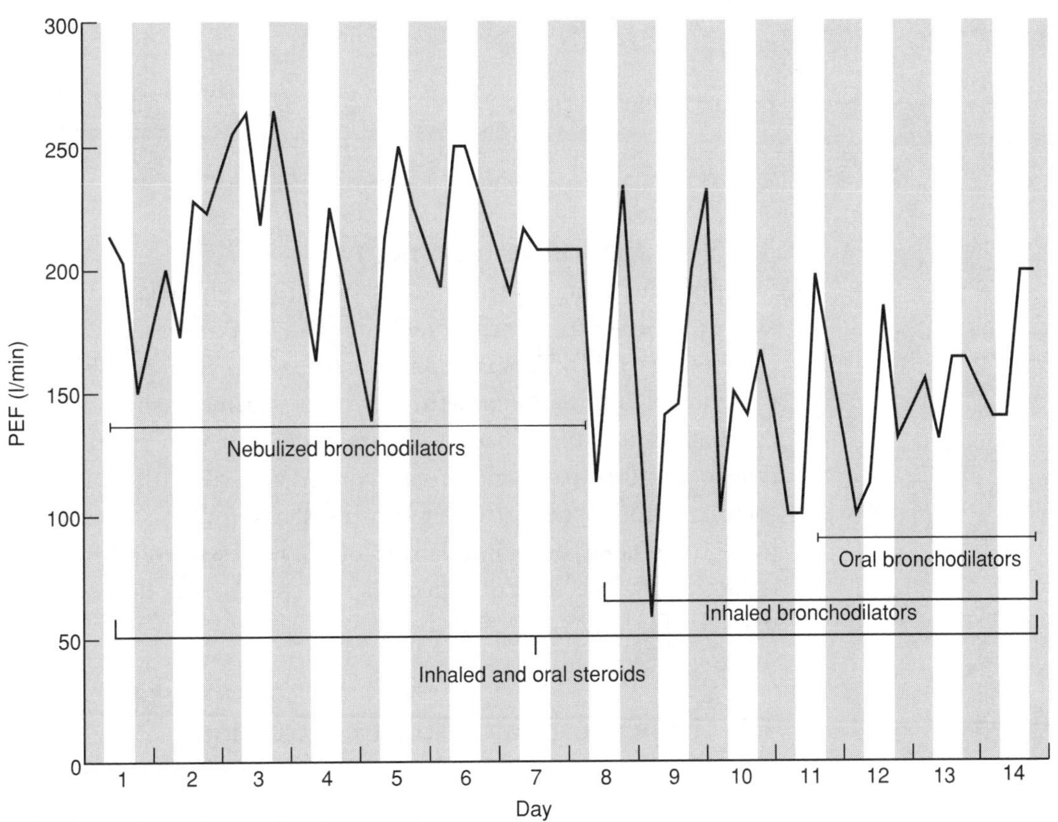

Figure 3.28 Loss of control of asthma when patient was switched from nebulized to inhaled bronchodilators.

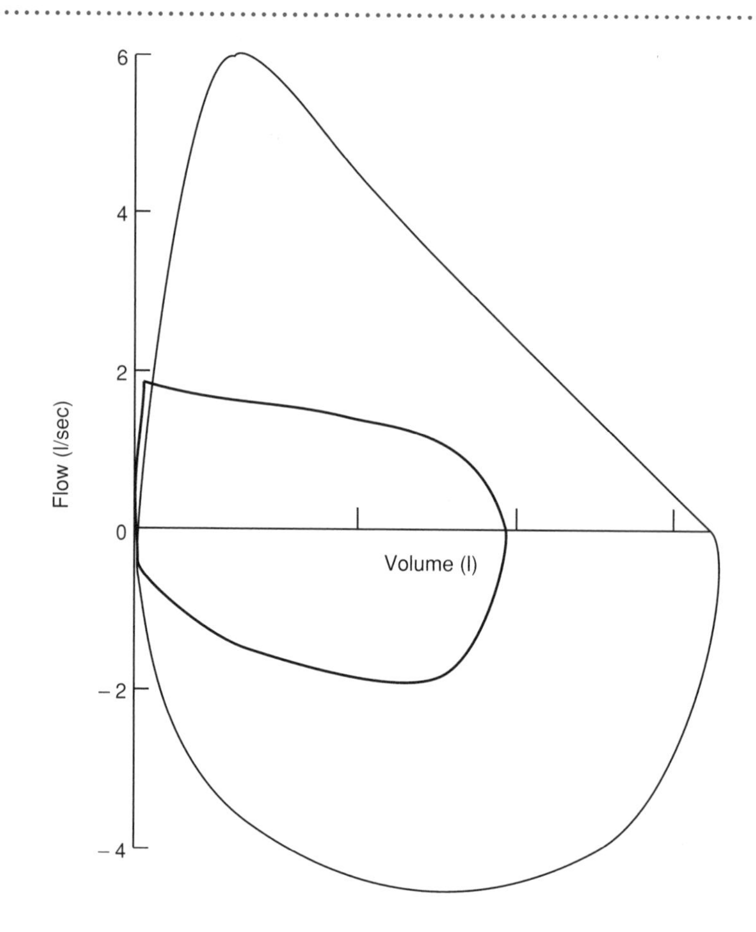

Figure 3.29 Upper airway obstruction in a young man with a tracheal tumour. FEV_1 = 1560 ml; PEF = 115 l/min; Empey Index = 13.5. Flow volume loop showed characteristic flattening (outer loop represents predicted shape). Conventionally, flow is always recorded in litres per second on a flow volume loop (rather than litres per minute).

3.6.5 Variation in treatment

Stopping and starting treatment without the physician or nurse being aware can lead to significant changes in peak flows for no apparent reason (Fig. 3.27). This is also important when trying to determine the presence of occupational asthma where treatment should be kept as constant as possible (Chapter 4). The other situation where a change in treatment needs to be monitored particularly carefully is in the recovery phase from acute asthma when transferring the patient from nebulized to inhaled bronchodilators, often representing a near ten-fold reduction in dose. Loss of control as reflected in peak flows can be quite marked (Fig. 3.28).

3.7 Upper respiratory tract obstruction

Peak flow can contribute to the diagnosis of upper airway obstruction. Where the upper airway is narrowed, e.g. due to a tracheal stricture or a goitre, peak expiratory flow will be reduced. However, so will the FEV_1, but to a lesser degree. The ratio of the FEV_1 (in ml) to peak expiratory flow (l/min) is called the Empey Index. A value of greater than 10 supports the diagnosis of upper airway obstruction, though values slightly below 10 may still be associated with significant tracheal narrowing.

The best test to confirm the diagnosis is the flow volume loop, where flattening of the expiratory peak and the inspiratory limb of the loop provides a characteristic picture (Fig. 3.29).

References and further reading

Dawkins, K.D. and Muers, M.F. (1981) Diurnal variation in airflow obstruction in chronic bronchitis. *Thorax*, **36**, 618–21.

Fishman, A.P. (1989) Cardiac asthma – a fresh look at an old wheeze. *N. Engl. J. Med.*, **320**, 1346–8.

Hope, J. (1835) *A Treatise on the Disease of the Heart and Great Vessels*, 2nd edn. W. Kidd, London, pp. 345–64.

Venables, K.M., Davison, A.G., Browne, K. and Newman-Taylor, A.J. (1989) Pseudo-occupational asthma. *Thorax*, **44**, 760–1.

4 Occupational asthma

> a breath thou art,
> Servile to all the skyey influences
> That dost this habitation, where thou keepst,
> Hourly afflict.
>
> William Shakespeare, *Measure for Measure*, III: i

Asthma caused by a specific agent in the workplace is called occupational asthma, and there are now over 200 documented causes (see Appendix F). When faced with a new case of asthma it is important to appreciate that this list is probably not complete and that you may have an 'index case' before you. If you believe a patient to have occupational asthma then referral to a chest physician is essential since confirming the diagnosis is not straightforward. Many of the causes are eligible for compensation under industrial accident rules in the UK (see Appendix F).

The proportion of a given workforce affected by a particular process may be very high. In the 1960s following the development of proteolytic enzymes in the detergent industry there were reports that 40–50% of workers developed occupational asthma. More recently in a study of asthma caused by azodicarbonamide, a compound used in the manufacture of polyvinylchloride and expanded foam plastics, 38% of workers had attended their doctor with symptoms but only 12% of cases were recognized as occupational asthma, and this after repeated attacks.

Delay in diagnosis and a prolonged exposure to a causative agent may lead to persistent asthma. Many patients continue to experience asthma of varying degrees years after apparent removal from the initial sensitizing agent.

A diagnosis of occupational asthma is reached by the combination of a good history and lung function testing.

4.1 History

The two most important indicators in the history when seeking a diagnosis of occupational asthma are improvement in symptoms on days away from work, and/or when on holiday. If patients improve on holiday it is important to define whether the holiday is at home, or away from home, since there may be a domestic cause for their asthma such as a newly acquired pet or a relevant hobby. In more severe occupational asthma it may take longer than the two days of the weekend before improvement is felt.

An occupational history must include specific details about the nature of the patient's occupation both current and past.

Duration of exposure is of relevance. The interval before a worker is sensitized may range from hours to years. If it is years before asthma develops, as is common for instance in platinum refining, then an occupational cause could easily be missed. Accidental exposure to high concentrations of irritant fumes such as smoke, ammonia, or acids may cause asthma which may last for years after a single exposure.

Once referred it is important for the respiratory physician to establish details of environmental influences in the factory such as type of ventilation, use of face masks and any changes in symptoms with climatic conditions.

Smoking and atopy have been shown to be predisposing factors in certain forms of occupational asthma. A mild asthmatic requiring infrequent treatment may develop more troublesome asthma if an occupational agent becomes involved. Such a case is as important as a new asthmatic with no previous history.

4.2 Objective assessment

Peak flow is the most practical method for establishing the presence of occupational asthma, and has been shown to have high sensitivity and specificity in the diagnosis.

The patient is carefully instructed to make readings every two hours from waking in the morning until going to bed at night. This should include at least one week at work followed by at least ten days off work and then two weeks at work again. Recording should continue over

	Monday	Tuesday	Wednesday	Thursday	Friday	Saturday	Sunday
Date	2/9	3/9	4/9	5/9	6/9	7/9	8/9
Time waking	7am	6.30am	7am	6.30am	7.30am	6.30am	8am
Time starting work	9am	9am	9am	9am	9am	9am	No work
Time stopping work	6pm	6pm	6pm	6pm	6pm	6pm	--
Time going to bed	10.30pm	11pm	10pm	11pm	10.50pm	12am	12.30am
Jobs done	Shop-keeping	Shop-keeping	Shop-keeping	Shop-keeping	Shop-keeping	Shop-keeping	Cooking
Treatment with Times	Inhaler 8am 1pm 5pm	Inhaler 7.30am 12.30pm 4pm 9pm	Inhaler 8am 1pm 5pm	Inhaler 8.30am 1.30pm 5.30pm	Inhaler 9am 11.30am 3pm 10pm	Inhaler 7.30am 12.30pm 4.30pm 9pm	Inhaler 10am 5pm
01.00 a.m.							
02.00 a.m.							
03.00 a.m.							
04.00 a.m.							
05.00 a.m.							
06.00 a.m.							
07.00 a.m.							
08.00 a.m.	380	360	340	350	340	350	
09.00 a.m.							380
10.00 a.m.	380	360	340	340	340	330	
11.00 a.m.							370
12.00 Noon	360	340	350	310	310	320	
01.00 p.m.	370						380
02.00 p.m.		320	360	330	300	340	
03.00 p.m.	350						390
04.00 p.m.		310	330	290	320	300	
05.00 p.m.	320						370
06.00 p.m.		300	310	350	350	340	
07.00 p.m.	340						
08.00 p.m.		320	340	360	360	320	360
09.00 p.m.	340						
10.00 p.m.	330	340	350	350	340	350	380
11.00 p.m.							
12.00a.m.						360	
MEAN	352	331	340	335	332.5	334	376

Figure 4.1 A filled record card. A 'day' in occupational asthma is defined as starting at the time of arrival at work and continues until the last reading before return to work the following day. This is because the early morning readings are likely to be a reflection of exposure to agents and treatment on the previous day. The shaded area shows the range of values for one 'day' showing a maximum value of 380, a minimum value of 320 and a mean value of 352. The first value on waking is included in the previous day's calculation. The data are then analysed in graphical form, mean daily peak flow and the maximum and minimum values attained in the 'day' being recorded on a daily plot.

weekends, and an attempt should be made to keep the environment of the factory and exposure to the agent as 'usual' as possible. Inhaled preventative therapy should be kept constant and a record of daily relief bronchodilator use kept. The jobs done and time of arrival at and the departure from work should be recorded daily.

> A 45-year-old woman presented with a history of 3 years' persistent but variable breathlessness. Occasionally she would be woken from sleep by wheezy breathlessness and initial peak flow monitoring showed a diurnal variation of up to 20%, suggesting a diagnosis of asthma. She was a life-long non-smoker but had worked for a number of years in the clothing industry, being involved in handling cloth sent from abroad for making into garments. When questioned she admitted that her symptoms improved slightly on a Sunday but much more so if she was away from the clothing shop for longer periods. A diagnosis of occupational asthma was entertained and occupational peak flow records obtained (Fig. 4.1).

4.2.1 Overall analysis of recording

Occupational peak flow records are analysed by dividing them into:

1. working weeks which show deterioration;
2. rest days which show improvement.

In the case example (Fig. 4.2), peak flow fell during all four exposure periods and rose in three out of four (75%) unexposed periods.

If 75% or more of such working weeks and rest days show these occupational changes then the individual has occupational asthma, and thus the diagnosis was confirmed in this patient. Such peak flow

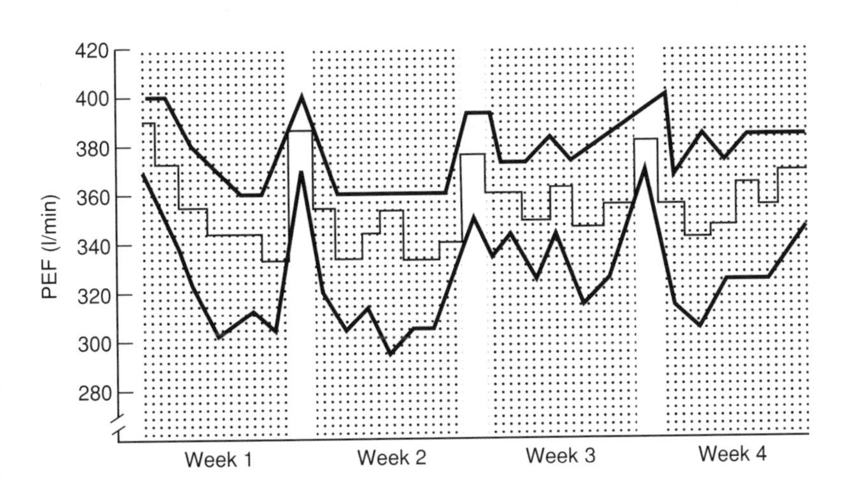

Figure 4.2 Plot of the raw data from Fig. 4.1 (Week 2) with data from the three other weeks. The stippled areas represent time spent at the clothing shop, the unstippled areas representing time away from work. The upper line represents daily maximum peak flow, the lowest the daily minimum, whilst the middle line is the mean value for each 24-hour period.

findings are a firm pointer to an occupational cause for asthma. If, as in this case, the person is exposed to a wide variety of incriminating substances, then further assessment by bronchial provocation tests may be necessary.

When interpreting these graphical patterns, there is usually good agreement between experienced observers. In one study complete agreement between four observers was found in 70%, with agreement between three out of four observers in 97% of cases. Therefore collaboration between observers will increase the sensitivity of the test.

4.3 Patterns of peak flow in occupational asthma

The patterns depend largely on the rate of recovery from exposure to the agent and the effect of repeated exposures. Patterns can show considerable differences between people exposed to the same agent, and even for the same individual in different weeks. Conversely, certain agents have a tendency to produce characteristic peak flow patterns.

Changes within one day Any variation in peak flow caused by occupational exposure is superimposed on the natural diurnal variation of peak flow.

(a) An **immediate** reaction to exposure to an agent gives rise to a record as shown in Fig. 4.3, with a drop in peak flow within minutes of exposure. However, exposure may continue at work throughout the day and then may take hours to recover.

(b) A **non-immediate** or late reaction occurs in cases where exposure to the agent results in a reaction some hours later. The peak flow response depends to a large extent on the time of the shift worked (Fig. 4.4). In a **morning shift** peak flow may continue to rise and be higher than on arrival at work before falling later in the day. The reaction may then occur after leaving work. In an **afternoon shift**, symptoms may be experienced earlier after exposure than in a morning shift since the diurnal variation is already naturally falling and a late asthmatic reaction exaggerates this fall. In a **night shift** worker the pattern of peak flow largely depends on the interval between waking and starting work.

(c) A flat record may be seen after repeated exposure to the agent, when peak flow may not recover by the following day, resulting in

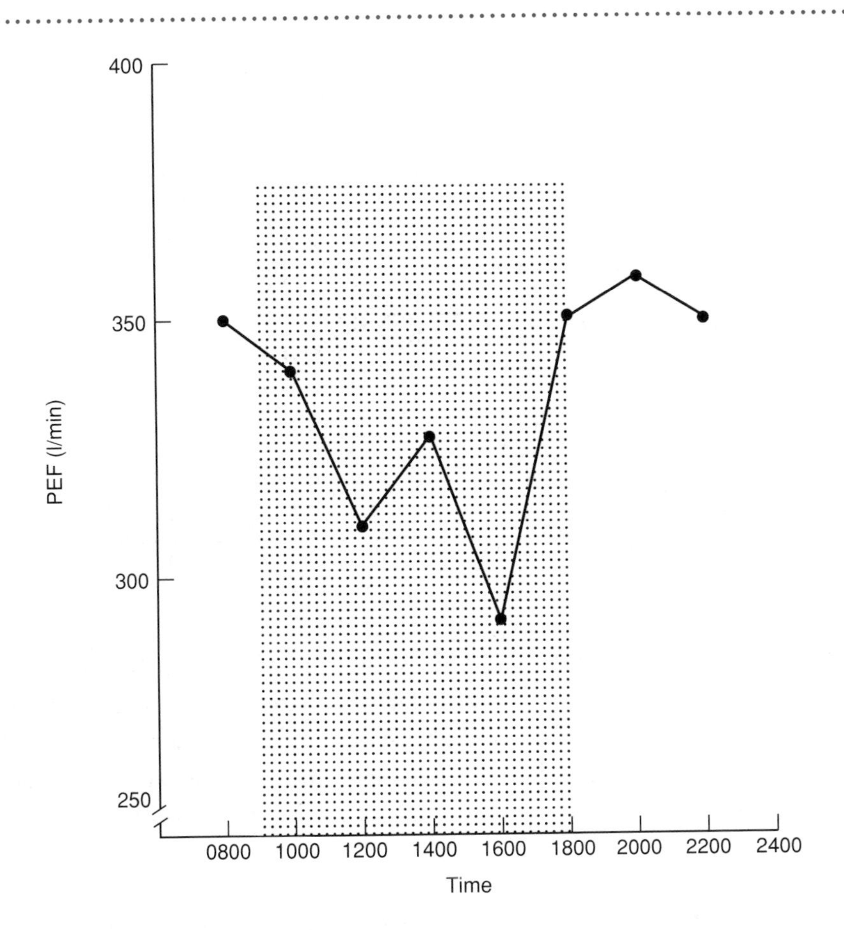

Figure 4.3 Immediate asthmatic reaction using data from Day 4 of Week 2 of the worked example. Stippled area represents exposure.

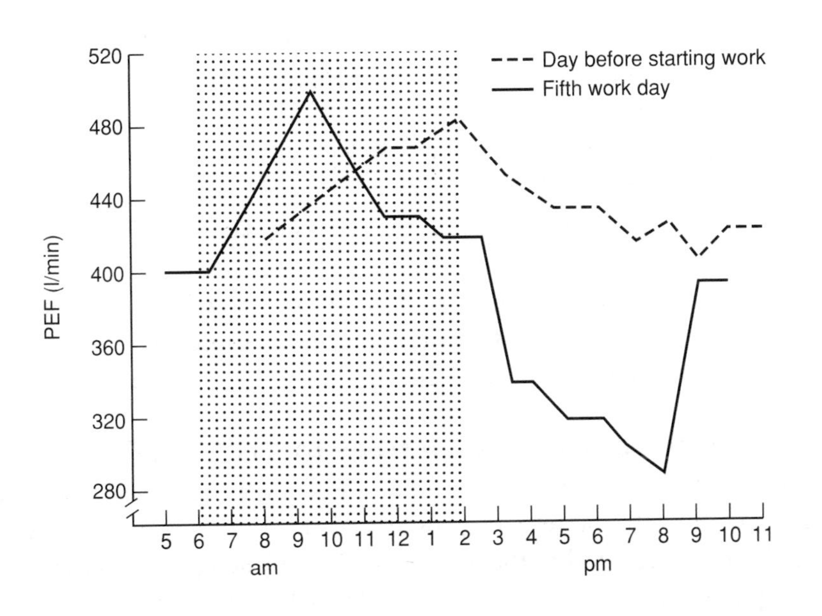

Figure 4.4 Record of hourly peak flow in a gravure printer sensitive to isocyanate fumes liberated from a laminating machine in a neighbouring room. The day before work shows a normal diurnal variation; the fifth working day shows a pronounced late asthmatic reaction starting after leaving work. (Redrawn from Burge (1982a), *Eur. J. Respir. Dis.*, **63** (Suppl. 123), 47–59 and reproduced with permission.)

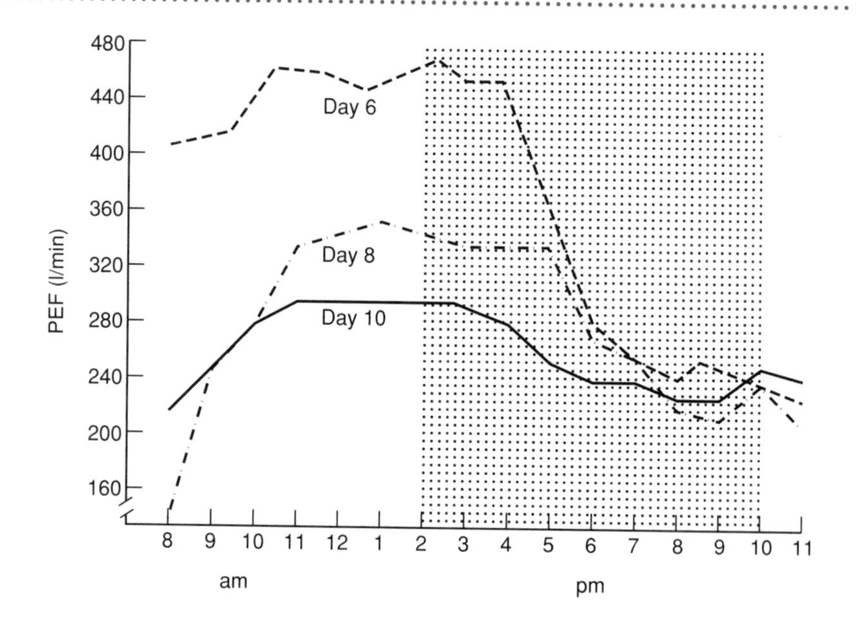

Figure 4.5 Record of hourly peak flow in the same gravure printer. Stippled area shows work exposure on an afternoon shift. A 'late' asthmatic reaction starts within two hours of going to work, day 10 shows a 'flat' pattern with little evidence of an hourly work-related reaction, due to lack of time for recovery between exposures. (Redrawn from Burge (1982a), *Eur. J. Respir. Dis.*, 63 (Suppl. 123), 47–59 and reproduced with permission.)

little diurnal variability. This is sometimes erroneously diagnosed as chronic bronchitis, particularly if there is little improvement with bronchodilators. The diagnosis is only likely to be made if the patient's peak flow can be shown to recover away from work, which in this type of case may take a long period of time (Fig. 4.5).

Changes in peak flow from day to day within the week

(a) Following a single exposure to an occupational allergen or irritant, morning dipping may continue for several days subsequently (Fig. 4.6).
(b) Equivalent deterioration on each day of exposure, with recovery at the weekend (within one day).
(c) Greater deterioration with each day of exposure (Figs 4.7 and 4.8).
(d) In some cases greater deterioration in peak flow is seen on the first day at work with lesser deterioration on subsequent days. This pattern needs to be convincingly shown not to be factitious.

Changes in peak flow from week to week This is a prominent feature in occupational asthma caused by wood dust or isocyanates, and gives a pattern of progressive deterioration until either a steady state is reached or the worker is forced by symptoms to leave the workplace until recovery has occurred. In such cases the recovery period

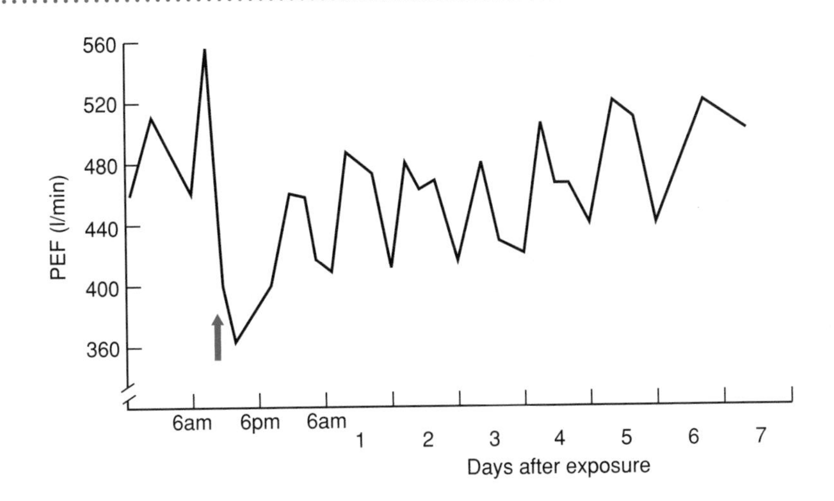

Figure 4.6 Four-hourly peak flow readings before and after a single exposure to tetrazine in a detonator manufacturer. The late reaction continues with morning dipping for at least seven days, with the patient experiencing symptoms of 'bronchitis', rhinitis and dermatitis. (Redrawn from Burge *et al.* (1984), *Thorax*, **39**, 470–1 and reproduced with permission.)

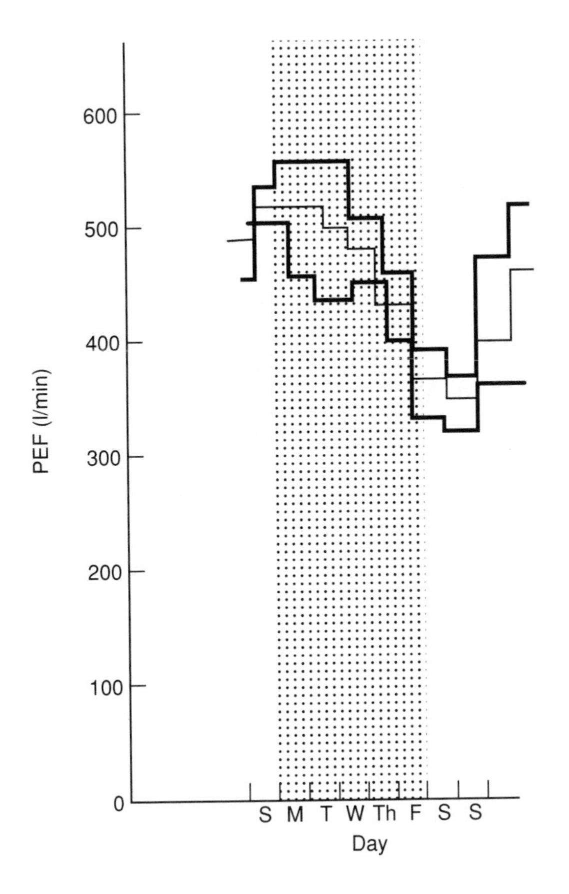

Figure 4.7 A 58-year-old manager in the timber industry showing repeated daily falls in peak flow following each exposure to Western Red Cedar dust over a period of 5 days with recovery over 2 subsequent days away from exposure.

is longer than 3 days. This means that recovery has not fully occurred after a 2-day weekend break and so the peak flow at the beginning of the week is still a reflection of exposure to agents the previous week (Fig. 4.9).

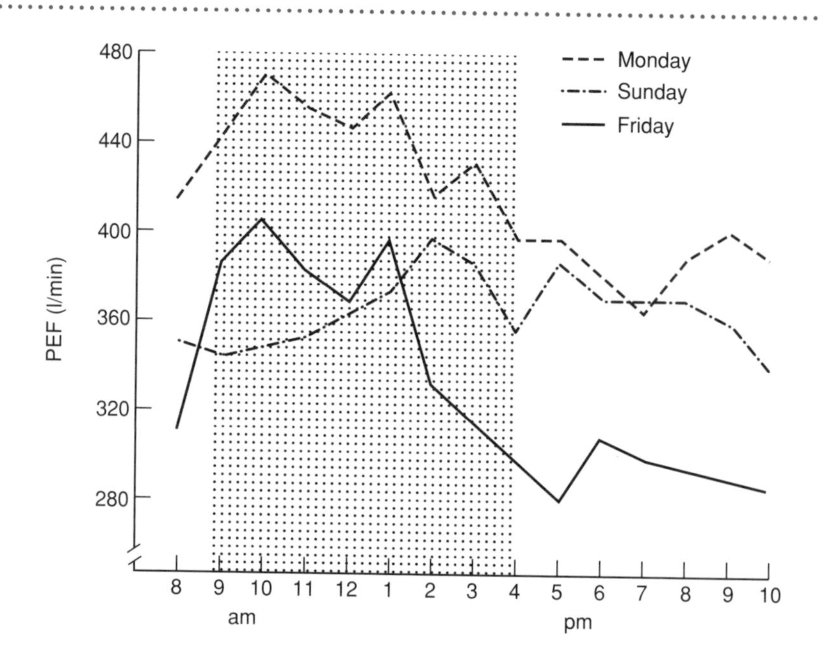

Figure 4.8 Plot of hourly peak flows in a production line worker in situ exposed to colophony. Sunday record at home shows a morning dip and exaggerated 'normal' diurnal variation. Morning dip has lessened by Monday morning but a later asthmatic reaction starts just before leaving work. By Friday of the same working week, morning dip and late asthmatic reactions are pronounced. This record shows progressive deterioration with work exposure. (Redrawn from Burge *et al.* (1979a), *Thorax*, **34**, 308–16 and reproduced with permission.)

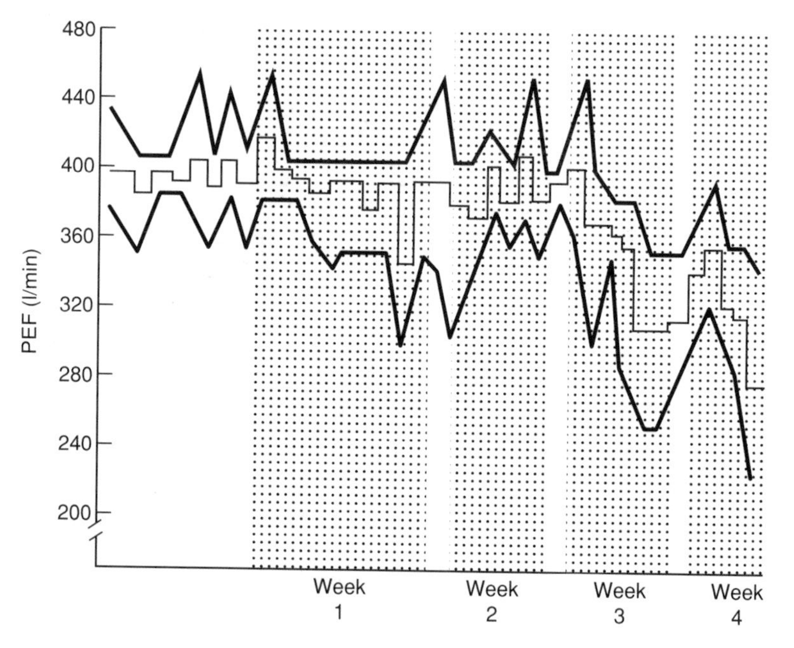

Figure 4.9 Record of daily maximum, mean and minimum peak flow in an inkmaker exposed to isocyanate fumes from a neighbouring factory. Deterioration started in the third working week and progressed further in the fourth working week. He was admitted to hospital with severe asthma three weeks later. (Redrawn from Burge *et al.* (1979a), *Thorax*, **34**, 317–23 and reproduced with permission.)

Maximum deterioration on the first day of the week with recovery on subsequent days Peak flow that is worst on the first day of the week but recovers on subsequent days is an infrequent feature in occupational asthma, occurring characteristically in byssinosis (Fig. 4.10), humidifier fever, polymer fume fever and grain fever.

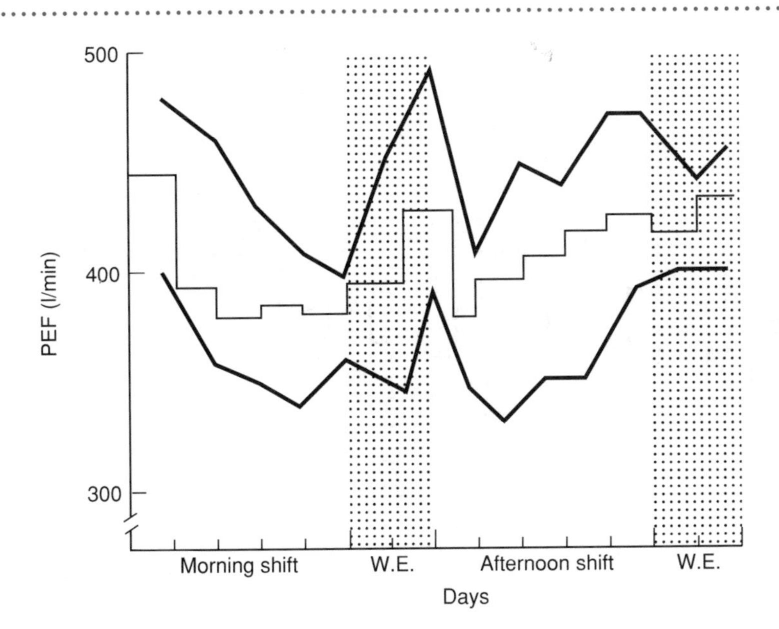

Figure 4.10 Byssinosis is a condition caused by inhalation of the dust of cotton, flax, hemp and sisal. It gives a non-immediate reaction which is worse at the beginning of the working week in its initial stages, and an increased severity of symptoms on the first working day after a prolonged break away from work than if the rest has only been a weekend. Usually at least 5 years' exposure to the dust is necessary before sensitization develops and this is as prevalent in atopic as non-atopic factory workers. As the symptoms worsen the asthma may progress to other days of the week but this progression is not invariable.

4.4 Pitfalls in interpretation of peak flow

In some cases the interpretation of peak flow is complicated by additional factors.

(a) Variable exposure to the occupational agent: which may be inhaled only occasionally but may cause symptoms and effects on peak flow records for a considerable time. The use of a protective mask can obliterate a work-related fall in peak flow (Fig. 4.11).

(b) Working shift patterns: care must be taken in adjusting 'days' when a worker changes the time of his/her shift.

(c) Altered waking times: on rest days the worker may get up later and so 'miss' an early morning dip of peak flow, hence the importance of stating wakening time on the chart (section 3.6.4, Fig. 3.26).

(d) Falsification of charts: occupational peak flows are particularly liable to forgery (section 2.7).

(e) Effects of co-factors. Climatic changes such as changes in humidity can affect responses to a particular substance (Fig. 4.12). Exposures will also vary according to the degree of environmental ventilation or whether the patient was exercising. A further complicating factor is that patients show a tendency to reduce treat-

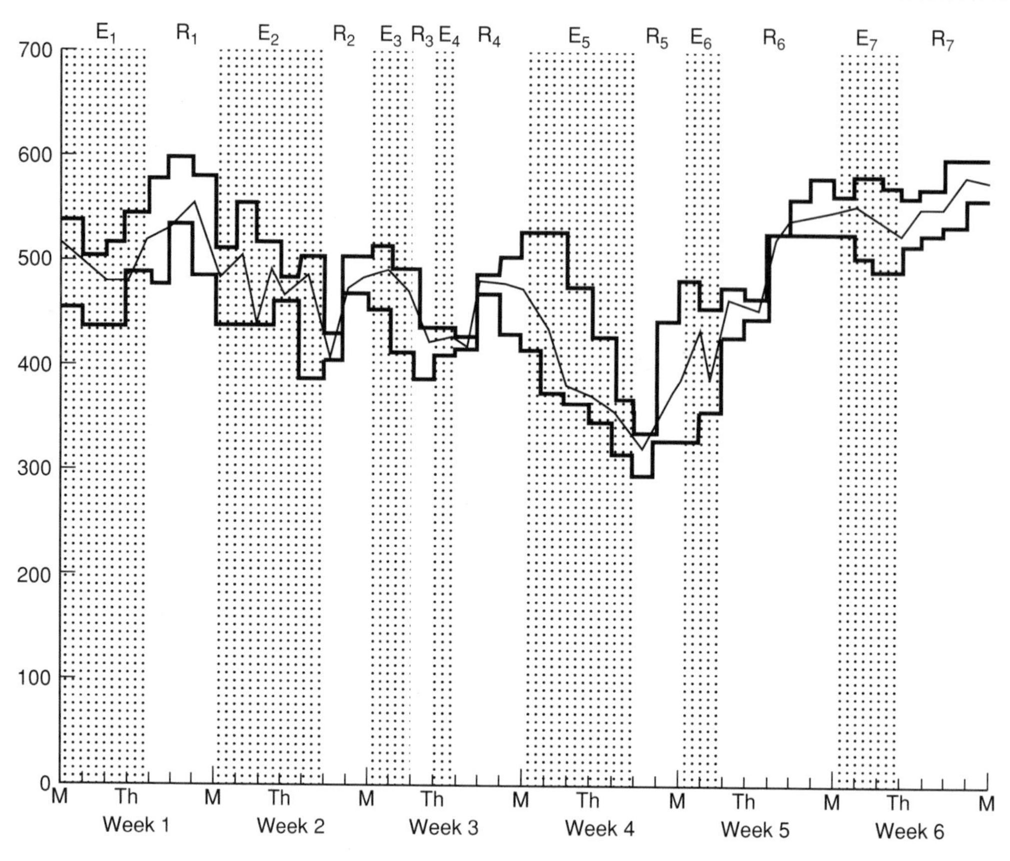

Figure 4.11 The same patient as in Fig. 4.7 with seven exposure periods (E1 to E7) to Western Red Cedar dust, followed by seven recovery periods (R1 to R7) of different length. There is deterioration in peak flow in exposure periods E2 to E6 with recovery in periods R1, R2, R4, R5 and R6. One day without exposure (R3) did not allow recovery. In exposure period E7 the patient wore a mask thus attenuating the fall in peak flow permitting maintenance of peak flow at recovery level achieved in period R6.

ment on days when symptoms are less troublesome; this confuses interpretation of work-related changes (Fig. 4.13).

Treatment with bronchodilators is too short-lived to have much effect on occupational peak flows but they should be taken **regularly** if taken at all to standardize the dose. There is a reduction in sensitivity of peak flows in the diagnosis of occupational asthma if inhaled/oral steroids or sodium cromoglycate is being taken. A study of workers exposed to colophony showed sensitivity was reduced from 77% in workers on no prophylactic medication to 42% in workers taking sodium cromoglycate or corticosteroids.

Although it is a cheap and non-invasive test, and does not involve taking any time off work for hospital admission, as would be necessary for bronchial provocation testing, peak flow has its limitations in occupational asthma. It may be dangerous for a severe asthmatic to return to the working environment for peak flow monitoring. In some

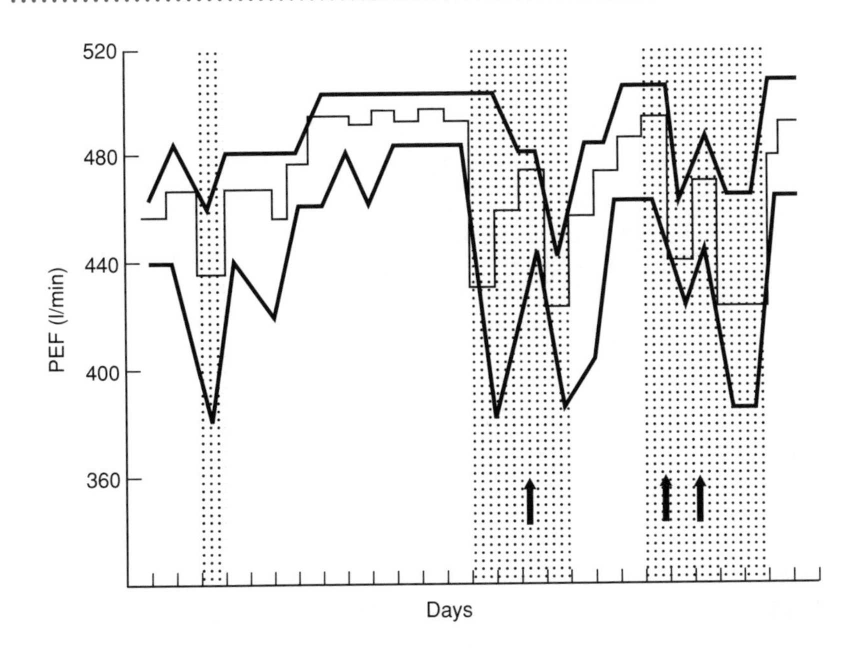

Figure 4.12 The maximum, mean and minimum peak flow in a worker using cyanoacrylate adhesive. Her symptoms are worsened on days when humidity is above 55% (↑), when more acrylic monomer becomes volatile. (Stippled areas represent days at work.) (Redrawn from Burge (1982a), *Eur. J. Respir. Dis.*, **63** (Suppl. 123), 47–59 and reproduced with permission.)

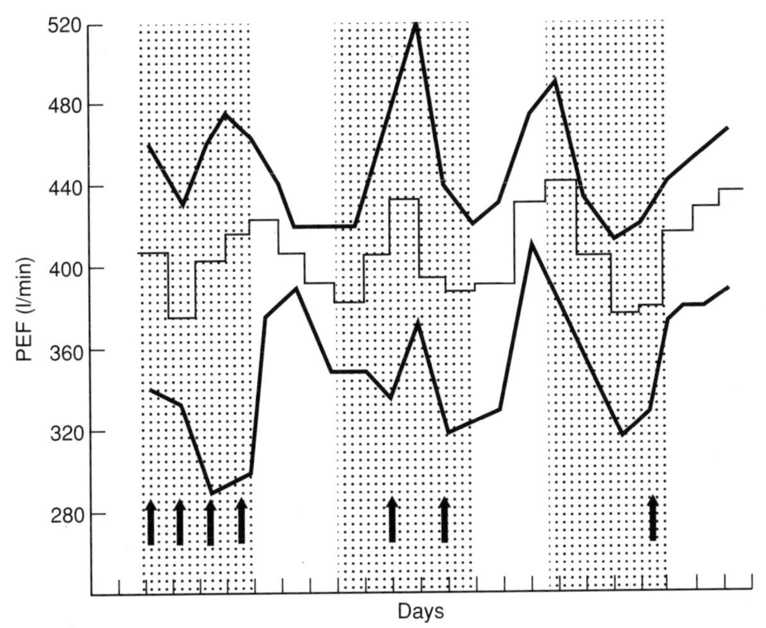

Figure 4.13 Daily maximum, mean and minimum peak flow in a laboratory animal technician who took salbutamol (↑) only on days at work (stippled areas), and who took cromoglycate during the second working week and the following weekend. The record shows an irregular pattern of mean peak flow, but the maximum peak flow is reduced (suggesting less bronchodilator use) and the minimum increased during days off work (suggesting recovery from an exposure). (Redrawn from Burge (1992a), *Eur. J. Respir. Dis.*, **63** (Suppl. 1), 47–59 and reproduced with permission.)

individuals symptoms will have led the worker to give up the job, and in these cases serial peak flows in the workplace cannot be used to clinch the diagnosis. Motivation to take the frequent readings required is considerable and it is not known how often this leads to inaccuracy.

References and further reading

Burge, P.S. (1982a) Single and serial measurements of lung function in the diagnosis of occupational asthma. *Eur. J. Respir. Dis.*, **63** (Suppl. 123), 47–59.

Burge, P.S. (1982b) Occupational asthma in electronic workers caused by colophony fumes: follow-up of affected workers. *Thorax*, **37**, 348–53.

Burge, P.S. (1984) Occupational asthma, rhinitis and alveolitis due to colophony. *Clin. Immunol. Allergy*, 4, 55–81.

Burge, P.S. (1987) Problems in the diagnosis of occupational asthma. *Br. J. Dis. Chest*, **81**(2), 105–15.

Burge, P.S., Hendy, M. and Hodgson, E.S. (1984) Occupational asthma, rhinitis and dermatitis due to tetrazene in a detonator manufacturer. *Thorax*, **39**, 470–1.

Burge, P.S., O'Brien, I.M. and Harries, M.G. (1979a) Peak flow meter records in the diagnosis of occupational asthma due to colophony. *Thorax*, **34**, 308–16.

Burge, P.S., O'Brien, I.M. and Harries, M.G. (1979b) Peak flow rate records in the diagnosis of occupational asthma due to isocyanates. *Thorax*, **34**, 317–23.

Chan Yeung, M., Lam, S. and Loener, S. (1982) Clinical features and natural history of occupational asthma due to Western Red Cedar (*Thuja plicata*). *Am. J. Med.*, 72, 411–15.

Cote, J., Kennedy, S. and Chan Yeung, M. (1990) Sensitivity and specificity of PC_{20} and peak expiratory flow rate in cedar asthma. *J. Allergy Clin. Immunol.*, **85**, 592–8.

Hudson, P., Cartier, A., Pinean, L. *et al.* (1985) Follow up of occupational asthma caused by crab and various agents. *J. Allergy. Clin. Immunol.*, **76**, 682–8.

Locewicz, S., Assoufi, B.K., Hawkins, R. and Newman-Taylor, A.J. (1987) Outcome of asthma induced by isocyanates. *Br. J. Dis. Chest*, **81**, 14–21.

Mitchell, C.A. and Gandevia, B. (1971) Respiratory symptoms and skin reactivity in workers exposed to proteolytic enzymes in the detergent industry. *Am. Rev. Respir. Dis.*, **104**, 1–12.

Newhouse, M.L., Tagg, B., Pocock, S.J. and McEwan, A.C. (1970) An epidemiological study of workers producing enzyme washing powders. *Lancet*, i, 689–93.

Paggiaro, P.L., Loi, A.M., Rossi, O. *et al.* (1984) Follow up study of patients with respiratory disease due to toluene diisocyante (TDI). *Clin. Allergy*, 14, 463–9.

Parkes, W.R. (ed.) (1994) *Occupational Lung Disorders*, 3rd edn. Butterworth-Heinemann, Oxford.

Perks, W.H., Burge, P.S., Rehahn, M. and Green, M. (1979) Work-related respiratory disease in employees leaving an electronics factory. *Thorax*, 34, 19–22.

Venables, K.M., Burge, P.S., Davison, A.G. and Newman-Taylor, A.J. (1984) Peak flow rate records in surveys: reproducibility of observer's reports. *Thorax*, 39, 828–32.

5 Management plans

> 'Thquire!' said Mr Sleary, who was troubled with asthma, and whose breath came far too thick and heavy for the letter s, 'Your thervant! Thith ith a bad piethe of bithnith, thith ith.'
>
> Charles Dickens, *Hard Times*, Book I, ch. 6

A peak flow meter prescribed without adequate instructions for management will soon be forgotten and become useless, and so it is essential that the patient must understand the need for treatment and know how to recognize and respond to an exacerbation. Having gauged the degree of severity of their asthma, a clear plan will be a guide to appropriate action.

This educational component requires a considerable amount of time but should reap dividends by reducing crises. Although studies giving general educational advice have not been shown to reduce asthma morbidity, more formalized 'management plans' have been devised and have been shown to reduce the use of oral steroids and the need for hospital admission.

The person best suited to aid the general practitioner in asthma education is a trained practice nurse who can ensure the patient has instructions on the correct use of the peak flow meter, inhalers and their function, and who will explain a management plan and above all answer questions.

The timing of instruction in self-management is important – it is unlikely to sink in when the patient is struggling with severe acute illness; shortly after recovery may be the best time to introduce these ideas since the problem will be a high priority for the patient. The British Thoracic Society has published guidelines on the management of acute and chronic persistent asthma and their principles should form the basis of a management plan.

5.1 Self-management based on peak flow

In this type of management plan the patient's best peak flow value is established for age, sex and height. This value may need to be modified if a higher value is obtained following oral or inhaled steroids, or lowered if there is a degree of irreversibility in the patient's airflow obstruction. It is also important to remember that a self-management plan based on peak flow values needs to be modified as a child grows in height and optimal peak flow increases.

Clear verbal and written instructions should be given as a guide to management at particular levels of peak flow, otherwise peak flow record cards may not be filled in when the patient becomes asymptomatic or acutely ill. Alternatively, values may be filled in while the patient monitors the onset of a crisis without realizing the significance of the observations. The patient should be told that in some individuals perception of airflow obstruction is poor and a peak flow meter may help detect an early deterioration in their asthma. The 'correct' percentage levels of optimal peak flow chosen to act as warning levels guiding treatment modification are still debated.

In the management plan originally devised by Beasley and colleagues for patients attending a hospital clinic, peak flow values at 70% and 50% of optimal peak flow (see Appendix H) are drawn across the peak flow chart and written into the self-management plan carried by the patient (Fig. 5.1). The 50% level was chosen because of the observation that morning dipping with a fall in peak flow of more than 50% from the optimal value often preceded sudden death from asthma in hospital.

The patient measures peak flow before inhaled bronchodilator use at least on wakening in the morning and more frequently if unstable.

- If the peak flow is greater than 70% of normal the patient continues on maintenance treatment, e.g. inhaled steroid twice daily.
- If the peak flow falls to under 70% but over 50% of normal over 36 hours, the dose of steroid inhaler is doubled and this dose maintained for the number of days required to achieve the previous baseline value and then for the same number of days again before returning to the previous maintenance dose.
- If the peak flow falls below 50% of normal prednisolone 40 mg daily (20 mg in children) is commenced and the general practitioner contacted (who will provide a replacement course in addition

Figure 5.1 Peak flow self-management plan. (*These values should be filled in by the nurse or physician when first giving the written plan to the patient.)

Target peak flow = •• *

If peak flow greater than 70% of normal
continue maintenance treatment:
(a) inhaled bronchodilator when needed
(b) inhaled steroid two times a day

If peak flow less than 70% of normal (•• *)
(a) double dose of inhaled steroid for •• * days required to achieve previous baseline
(b) continue on this increased dose for •• * days
(c) return to previous dose of maintenance treatment

If peak flow less than 50% of normal (•• *)
(a) start oral prednisolone 40 mg daily (20 mg daily for children) and contact your general practitioner
(b) continue on this dose for the number of days required to achieve previous baseline
(c) reduce oral prednisolone by 5 mg every other day to zero

If peak flow less than 30% of normal (•• *)
(a) contact general practitioner urgently or, if unavailable
(b) contact ambulance or, if unavailable
(c) go directly to hospital and if possible take bronchodilator via nebulizer before leaving house

to assessing the patient). Prednisolone is continued at this dose until the target peak flow is achieved. It may then be taken at half the initial dose for the same number of days again, or else slowly tailed off in 5 mg steps to zero.

- If the peak flow falls to less than 150–200 l/min then the general practitioner is contacted urgently or, if unavailable, an emergency ambulance is called. If neither of these can be obtained and nebulized bronchodilator is available at home then this should be taken first before the patient is taken to hospital.

The lower action point of 150 l/min may require modification in certain patients. Such a reading in a tall young rugby player would indicate a severe attack, but might be quite close to normal in a frail elderly woman. To some extent the self-management plan should be individualized and based on past experience. Apart from the somewhat arbitrary lower action point which may be too high or too low in some patients, a patient known to have rapid declines in peak flow may need to modify inhaled steroid treatment before peak flow falls to 70% of optimal. Many successful management plans used in general practice use 80% as the first intervention level. Furthermore, in the generally less severe asthmatics seen in general practice a 40% level for starting prednisolone may be more appropriate. A worked example is given in Fig. 5.2. Appendix H gives a table of percentages of optimal peak flows for quick identification, in a clinic, of the relevant percentage peak flow levels.

Name: John Smith
Target Peak Flow: 600 litres/minute
Peak flow to be measured: 4 times daily
Routine treatment: Beclomethasone 250 2 puffs twice daily
Salbutamol 2 puffs as required

1. If peak flow falls to between 420 and 300 consistently over 36–48 hours, or if you develop a 'cold'
(a) increase beclomethasone to 4 puffs twice daily until target peak flow is acheived
(b) continue at this dose for a further 5 days then return to original dose

2. If peak flow falls to less than 300
(a) start oral prednisolone 30mg daily and inform GP
(b) then stay at 30mg until target peak flow is achieved
(c) then reduce by 5mg daily to zero

3. If peak flow falls to less than 150
(a) contact GP urgently; or
(b) ring for ambulance; or
(c) go directly to hospital; and
(d) use nebulizer: salbutamol 5mg

Figure 5.2 Worked example of a management plan for asthma using peak flow readings.

Electronic methods of recording serial peak flows are now feasible; they overcome manual recording errors and are used quite extensively at present in research. Such machines are able to store 2–4 weeks of data on a clip which can be downloaded directly onto compatible software. Similar 'logging' meters capable of measuring FEV_1 are under development.

A 10-year-old boy presented with recurrent paroxysmal attacks of coughing. A lack of objective measurement in the form of peak flow led to difficulty in monitoring the adequacy of inhaled steroid therapy and finally resulted in a prolonged course of oral prednisolone.

He was then assessed by a trained practice nurse and provided with a self-management plan which resulted in achievement of above-maximal predicted peak flows and stable asthma. On two occasions a relapse was noted and appropriate action was taken by his mother. On the second occasion, by early doubling of his inhaled steroids a course of oral steroids was avoided.

At his last visit his mother commented that she felt the threshold for increasing inhaled steroids should be higher than 70% in his case. However he was able to run again and she could not believe that this was the unwell child she had known 6 months earlier. The educational input required to set up such a management plan is achieved with difficulty in a short 5–10 minute consultation (either in general practice or a hospital outpatient clinic). Protected time for this is best provided by a motivated and trained practice nurse with back-up from the doctor appropriate to the nurse's confidence and knowledge.

Table 5.1 sets out the various stages in this case history, from first presentation to the eventual achievement of a stable condition 15 months later.

Table 5.1 Case history showing benefit of a well-delivered management plan

Date	*Presentation*	*Peak flow*	*Diagnosis*	*Treatment*
29 01 91 (GP)	Dry cough, wheeze on modest exertion nocturnal coughing O/E chest clear	300 l/min (predicted for age, height 400 l/min)	Asthma	Beclomethasone rotacap 200 μg b.d. (400 μg/day), salbutamol rotacap 200 μg b.d.
07 02 91 Practice Nurse	Asthma clinic Asthma controlled on treatment	360 l/min		Re-taught inhaler technique
25 09 91 (GP)	Persistent barking cough. Has stopped his treatment O/E chest clear	300 l/min	Asthma	Recommenced on above treatment
02 10 91 (GP)	Cough, sore throat. O/E chest clear	None	Pharyngitis	Oral penicillin
07 10 91 (GP)	Cough worsening, especially at night	None	Upper respiratory infection	Erythromycin, cough linctus
10 10 91 (GP)	Cough persists night and day with some wheeze. O/E chest clear	340 l/min (afternoon surgery)	Uncontrolled asthma	Add prednisolone 20 mg
17 10 91 (GP)	Generally much better but coughing again today. Parents both smokers O/E chest clear.	None		Salbutamol rotacaps x 6/day Reduced prednisolone 15 mg
25 10 91 (GP)	'Back to normal'	None		Continue beclomethasone rotacaps 200 μg b.d.(400 μg/day), salbutamol rotacaps 200 μg p.r.n.
10 12 91 (GP)	Recurrence of persistent cough O/E chest clear	None	Uncontrolled asthma	Salbutamol rotacap 200 μg every 2 hr, beclomethasone rotacap 400 μg b.d. (800 μg/day), salbutamol CR 4 mg b.d., amoxycillin 125 mg t.d.s., prednisolone 15 mg/day
11 12 91 (GP)	Cough persists	None	Uncontrolled asthma	Continue treatment
13 12 91 (GP)	Coughing spasms day and night. Chest clear	None		Prednisolone 30 mg, switch to beclomethasone 100 inhaler ii b.d. (400 μg/day), salbutamol inhaler ii b.d., both via Volumatic
20 12 91 (GP)	Improving	None		Reduction to 20 mg prednisolone/day and tailed off over subsequent 2 weeks
02 01 92 Practice Nurse	Full assessment			Provided with peak flow meter and taught recording on charts

Table 5.1 (*Continued*)

Date	*Presentation*	*Peak flow*	*Diagnosis*	*Treatment*
16 01 92 Practice Nurse	Assessment of peak flow chart and symptoms			Management plan with best achieved (450 l/min) as target: 70% level 315 l/min 50% level 220 l/min
19 01 92 (GP)	Recurrent cough	250 l/min		Mother starts 30 mg prednisolone and contacts GP who advises to continue as in management plan
11 03 92 (GP)	Recurrent cough	300 l/min		Mother increases beclomethasone 100 inhaler to v b.d. (1000 μg) and hence avoids oral steroids
09 04 92 Practice Nurse	Reviewed 'well'	400 l/min		On beclomethasone 100 inhaler iii b.d. (600 μg/day)
28 04 92 Practice Nurse	Reviewed 'well'	430–450 l/min		Asthma continues stable

5.2 Zoned management plans

An extension of the self-management plan incorporates a coloured zone system with treatment interventions guided by peak flows. Levels should be set to suit individual circumstances and previous experience. In practice the levels at which action is taken are given as actual peak flow values rather than as percentages of predicted peak flow, which makes it easier for the patient to use.

The following is an example of a zone plan used in the United States (Fig. 5.3).

- **Green Zone**: peak flow is between 80 and 100% of personal best or per cent predicted and variability is less than 20%. There should be minimal asthma symptoms in this zone.
- **Yellow Zone**: peak flow is between 50 and 80% of predicted with 20–30% variability and/or symptoms of asthma. This is indicative of an exacerbation requiring a temporary increase in inhaled bronchodilator therapy and possibly oral corticosteroids. If there has been a gradual decline in peak flow and little response to

ASTHMA CONTROL PLAN FOR ______________________________
(name of patient)

PREPARED BY ______________________________, M.D.

Green Zone: All Clear

This is where you should be every day.

Peak flow between ________ (80-100% of personal best)*

No symptoms of an asthma episode. You are able to do your usual activities and sleep without having symptoms.

The doctor will check which applies to you.

☐ Take these medicines.

Medicine	*How much to take*	*When to take it*

☐ Follow your asthma trigger control plan to avoid things that bring on your asthma.

☐ Take ________________ (medicine) before exercise.

Yellow Zone: Caution

This is not where you should be every day. Take action to get your asthma under control.

Peak flow between ________ (50-80% of personal best)*

You may be coughing, wheezing, feel short of breath, or feel like your chest is tight. These symptoms may keep you from your usual activities or keep you from sleeping.

☐ ***First***, take this medicine:

Medicine	*How much to take*	*When to take it*

☐ ***Next***, if you feel better in 20 to 60 minutes and your peak flow is over ________ (70% of personal best), then:

☐ Take this medicine

Medicine	*How much to take*	*When to take it*

☐ Keep taking your green zone medicine(s).

☐ ***But***, if you DO NOT feel better in 20-60 minutes or your peak flow is under ________________ (70% of personal best), **follow the Red Zone Plan.**

Let the doctor know if you keep going into the Yellow Zone. Your Green Zone medicine may need to be changed to keep other episodes from starting.

Red Zone: Medical Alert

This is an emergency! Get help.

Peak flow under ________ (50% of personal best)*

You may be coughing, very short of breath, and/or the skin between your ribs and your neck may be pulled in tight. You may have trouble walking or talking. You may not be wheezing because not enough air can move out of your airways.

☐ ***First***, take this medicine:

Medicine	*How much to take*	*When to take it*

☐ ***Next*, call the doctor** to talk about what you should do next.

☐ ***But*, see the doctor RIGHT AWAY or go to the hospital if *any* of these things are happening:**

–Lips or fingernails are blue

–You are struggling to breathe

–You do not feel any better 20 to 30 minutes after taking the extra medicine and your peak flow is still under ________ (50% of personal best).

–Six hours after you take the extra medicine, you still need an inhaled $beta_2$-agonist medicine every 1 to 3 hours and your peak flow is under ________ (70% of personal best).

**This is a general guideline only. Some people have asthma that gets worse very fast. They may need to have a yellow zone at 90-100% of personal best.*

Figure 5.3 Example of a zoned management plan. (Reproduced with permission from International Consensus Report on the Diagnosis and Management of Asthma. *Clin. Exp. Allerg.* (1992), **22**, Suppl. 1, 16.)

inhaled bronchodilator then a course of oral corticosteroids is likely to be necessary until the peak flow returns to the Green Zone, and then discontinued. If the patient is already on inhaled corticosteroids then they should be doubled in dose until the peak flow improves. If the peak flow is frequently in the Yellow Zone then it is likely that treatment in the Green Zone needs modification and is inadequate.

- **Red Zone**: this indicates that urgent attention to the asthma is required. Peak flow is below 50% predicted. Asthma symptoms are present. If peak flow remains below 50% despite inhaled bronchodilator then emergency medical advice should be sought. If peak flow improves into the Yellow Zone then action in the latter should be followed. Falls into the Red Zone may signify that therapy in the Green Zone is inadequate.

Vitalograph have recently produced a peak flow meter (see Appendix A) to help in the use of this type of management plan, although this

Name: Age: Review date:

Symptoms only self management plan

- When you feel normal
  Continue maintenance treatment:
  (a) Bronchodilator two times a day or when needed
  (b) Inhaled steroid two times a day
- If you get a cold or start to feel tight
  Use your bronchodilator two puffs every 6 hours
- If you wake with wheezing at night or have a persistent cough
  (1) Double dose of inhaled steroid for number of days it takes you to return to normal and continue for a further work at this dose.
  (2) Use bronchodilator two puffs every 6 hours
- If your bronchodilator only lasts 2 hours and you find doing your normal activities makes you short of breath
  (1) Start oral prednisolone 40 mg daily (20 mg daily for children) and contact general practitioner
  (2) Continue to use this dose for the number of days required to return you to normal
  (3) Reduce oral prednisolone to 20 mg daily (10 mg daily for children) for same number of days
  (4) Stop prednisolone
- If your bronchodilator lasts only 30 minutes or you have difficulty talking, call the doctor immediately. If you are unable to contact the surgery take your inhaler and go by car to a hospital casualty department or call an ambulance

Figure 5.4 Symptoms only self-management plan, which may be used in patients for whom peak flow based management plans are inappropriate.

does not include a system for colour coding on the scale itself such as has been employed by Charlton. However these plans require further evaluation to ascertain effectiveness in controlling asthma.

5.3 'Symptoms-only' self-management plan

A 'symptoms-only' self-management plan for asthma has also been devised which does not involve the use of a peak flow meter. An example of the instructions in this plan is shown in Fig. 5.4.

The peak flow management plan has been used with success in both hospital outpatient departments and in general practice. The symptoms-only management plan has only been studied in less severe asthmatic patients in general practice and as yet it is not known whether it is as effective in more severe patients attending a hospital clinic. In the studies to date both types of plan have shown a significant reduction in oral steroid use, morbidity and doctor consultations for asthma.

Peak flow in children

6

'But wait a bit,' the Oysters cried,
'Before we have our chat;
For some of us are out of breath,
And all of us are fat!'

Lewis Carroll, *Through the Looking Glass*, ch. 4

6.1 Peak flow in diagnosis

Peak flow measurement in children suspected of having asthma is as helpful in making the diagnosis as it is in adults. Peak flow can be measured in most children over the age of 5 and, occasionally, younger than this. It is important when teaching the technique that a balance is achieved between getting the best results and losing the child's concentration, and it has been suggested that a maximum of five blows should be performed at each session.

It is important to be aware of pitfalls in the actual technique of blowing and to observe the child carefully during the procedure. The problems encountered are very similar to those seen in adults (section 2.1.2) but special attention should be paid to rule out the 'spitting' technique which may increase the peak flow by as much as 50 l/min. It is also very important to *avoid* measurement of peak flow in a child with diagnosed asthma until a reliable and effective inhaler technique has been well established. Getting the child to breathe in from their inhaler must take precedence over exhaling into a peak flow meter. Some children can do one of these but not both!

Once a reliable peak flow reading has been recorded in the clinic it can be used to make the diagnosis by the following measures.

1 Compare the child's peak flow reading with the expected value. In children peak flow relates to height but not age or sex (see Fig. 1.9).

2 Observe response to inhaled β_2-agonists or inhaled/oral steroids (Fig. 3.11).
3 Observe the response to a simple challenge such as the free running exercise test (Fig. 3.9).
4 Monitor peak flow at home twice daily over a few weeks to establish diurnal and day-to-day variability. This can be particularly useful where the presentation is less well recognized, for instance in the child whose symptom is recurrent cough (Fig. 3.2).

It is vital to establish a 'best' result for the individual child and, once established, to use this reading to compare other readings when he/she is unwell. Like most other physiological measurements there is a normal range which will change with increasing height.

Emphasis should not be placed too heavily on a 'one-off' reading in clinic, the dangers of which are highlighted in Chapter 2 (Fig. 2.5).

It is important to remember variability of readings between different peak flow meters (section 3.6.2) and to encourage children to bring their own 'home' peak flow meter to clinic to maintain continuity of readings.

Performing a peak flow may be a useful stepping stone towards measuring FEV_1, particularly for children with a diagnosis other than asthma in whom the peak flow measurement is not very useful. Spirometry, i.e. FEV_1, can usually be performed reliably by about 7 years of age, although as with peak flow measurements this is variable. Spirometry or a flow volume loop (Chapter 1) is an important measurement in restrictive lung disease (for example neuromuscular disease, scoliosis) and also in mixed obstructive/restrictive disease such as cystic fibrosis where bronchiectatic damage to the central airways can cause peak flow readings to underestimate the degree of lung damage. Spirometry and flow volume loops avoid this pitfall by being less effort dependent and measuring function in the small airways.

After the diagnosis of asthma has been made in a child, then measurement of peak flow should be used to monitor progress. In the majority of children this will be short term, until the asthma is stable. Regular continuous peak flow measurement and recording is indicated only in troublesome asthma where control is difficult, or in brittle asthma where sudden precipitous falls in peak flow heralding life-threatening attacks may occur (Fig. 3.21). Lastly, there are causes other than asthma of low peak flow measurements. These are very similar to those in adults (Table 1.2) but should also include disorders such as croup (acute laryngo-tracheo-bronchitis) and sub-glottic stenosis, often following prolonged intubation in the neonatal period.

6.2 Peak flow in school

Asthma is the commonest chronic disease encountered in children and results in about 20% of all days lost from school in the UK. In the recent General Household Survey it was the most frequently reported longstanding illness in children. A study in North Tyneside in 1983 showed that 11% of 7-year-old children had asthma, and higher figures than this have been reported from other areas. At least two or three children in a class of 20 are likely to have asthma. Young children tend to lose more time from school as a consequence of asthma than older children. This may lead to feelings of inferiority in the child, frequent short episodes of school loss tending to be more disruptive academically than an occasional longer period away from school.

The support that children with asthma obtain from their school depends to a large extent on their teachers' knowledge of the condition. School nurses are in a particularly good position to educate school staff; in a study of some inner London schools only 4% of teachers had received any training in asthma management. Some teacher training colleges now include education in asthma as part of the course.

A Nottingham study showed that only 10% of primary school children and 50% of secondary school children were allowed to keep and use their inhalers by themselves in school: 32% of children attending a hospital clinic had no access to inhalers at school. This was largely due to worries that the children might overdose or become addicted, that inhalers might be lost or used by other children. Most physicians feel that children should be responsible for managing their condition and be taught to avoid crises as early as possible.

Clearly there is room for improvement in communication between medical staff and teachers. The ability to use and interpret a peak flow meter should be most helpful in school. Most children can do a peak flow by the age of 5 years. Many asthmatics go unrecognized for long periods of time, learning to limit their physical activity in order to suffer less symptoms. A peak flow meter in an easily accessible place together with a set of normal values would be helpful in the assessment of an acute exacerbation and perhaps to the school nurse as a 'screen' of

Figure 6.1 The National Asthma Campaign School Asthma Card.

Card checked

Date	Initials

Providence House
Providence Place
London N1 0NT

Facsimile
0171 704 0740

If you would like a copy of the pamphlet *Asthma at school* please write to the address above.

4

NATIONAL **ASTHMA** CAMPAIGN
getting your breath back

School asthma card

Name

Address

Telephone — Home

Parent's work

General Practitioner — Name

Telephone

Consultant

Hospital — Name

Reference number

Telephone

This card is for your school. Remember to update it if treatment is changed.

1

Regular treatment to be taken in school time

Name and how taken	Dose and when taken

Before exercise

2

Relief treatment when needed

For sudden chest tightness, wheeze, breathlessness or cough, give or allow child to take

Name and how taken	Dose and when taken

If no relief or symptoms reappear within three hours

- Repeat above
- Call parent

If child is fighting for breath, speechless or blue

- Repeat above
- Call parent
- Dial 999 for an ambulance
  or
- take to nearest hospital

3

children with persistent respiratory symptoms where a low peak flow should lead to referral to the child's general practitioner. However, a normal peak flow in this situation does not rule out a diagnosis of asthma. Children with asthma should carry a school asthma card with clear instructions for management in an emergency, and precautions to be taken in particular situations. The National Asthma Campaign produce such a card (Fig. 6.1). Children with more severe asthma need serial peak flows, including values taken at lunchtime.

References and further reading

Ayres, J.G., Pansari, S., Weller, P.H. *et al.* (1992) A high incidence of asthma and respiratory symptoms in 4–11-year-old children. *Respir. Med.*, **86**, 403–8.

Bevis, M. and Taylor, B. (1990) What do school teachers know about asthma? *Arch. Dis. Child.*, **54**, 622–5.

Douglas, J.W.B. and Ross, J.M. (1965) The effects of absence on primary school performance. *Br. J. Educ. Psychol.*, **35**, 28–40.

Greenough, A., Everett, L. and Price, J. (1990) Are we recording peak flows properly in young children? *Eur. Respir. J.*, **3**, 1913–96.

Hennessy, E., Masson, J. and Russel, G. (1987) How readily do children with asthma have access to their treatment? *The Practitioner*, **231**, 1261–2.

Higgins, B.G., Britton, J.R., Chinn, S. *et al.* (1992) Comparison of bronchial reactivity and peak expiratory flow variability measurements for epidemiological studies. *Am. Rev. Respir. Dis.*, **145**, 588–93.

Hill, R.A., Britton, J.R. and Tattersfield, A.E. (1987) Management of asthma in schools. *Arch. Dis. Child.*, **62**, 414–15.

Lee, D.A., Winslow, N.R., Speight, A.N.P. and Hey, E.N. (1983) Prevalence and spectrum of asthma in childhood. *Br. Med. J.*, **286**, 1258–60.

Mitchell, R.G. and Dawson, B. (1973) Educational and social characteristics of children with asthma. *Arch. Dis. Child.*, **48**, 467–71.

Richards, W. (1986) Allergy, asthma, and school problems. *J. School Hlth*, **56**, 151–2.

Speight, A.N.P., Lee, D.A. and Hey, E.N. (1983) Underdiagnosis and undertreatment of asthma in childhood. *Br. Med. J.*, **286**, 1253–6.

Appendix A Peak flow meters

A.1 History

The nineteenth-century physician who instructed a patient with respiratory disease to blow out a candle was in effect assessing peak flow.

Peak flow was first used as a physiological measurement by Hadorn in 1942, measuring expiratory flow across a mouthpiece with an aneuroid manometer. He measured flows of up to 500 l/min, judging the momentary deflection of the pointer by eye (Fig. A.1) and producing results very comparable to modern peak flow meters (Fig. A.2). A variety of instruments was developed over subsequent years, culminating in the pneumotachograph, used subsequently for calibration of peak flow meters.

The original Wright peak flow meter was introduced in 1959. It followed work by Wright and McKerrow at the Medical Research Council Pneumoconiosis Unit in Penarth, South Glamorgan. Although the Wright peak flow meter does not require the electricity supply necessary for most previous instruments, its expense and relative bulk precludes its routine use for domiciliary peak flow measurement. This niche has been filled by various mini peak flow meters, which are durable, lightweight, cheap, easy to carry and simple to use.

The turbine spirometers are portable but more complex and expensive devices which measure FEV_1 and FVC in addition to peak flows.

Peak flow meters currently available are as follows:

Peak flow meters

Wright peak flow meter
Mini Wright peak flow meter
Vitalograph peak flow meter
Ferraris Pocketpeak peak flow meter

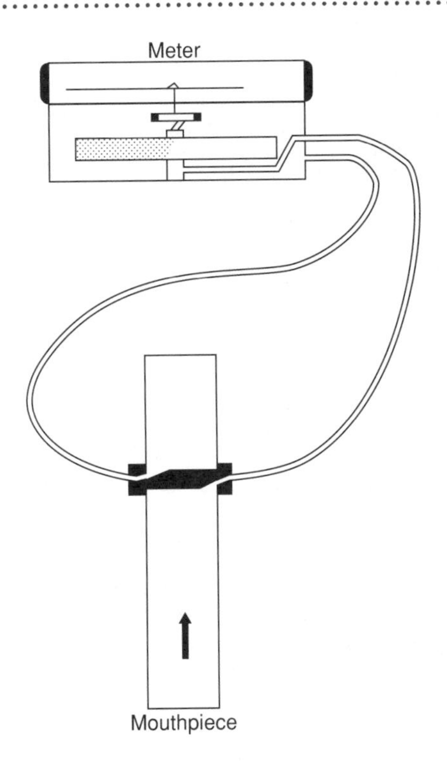

Figure A.1 Cross-section of the first aneroid meter for measuring airflow. (After Hadorn, 1942.)

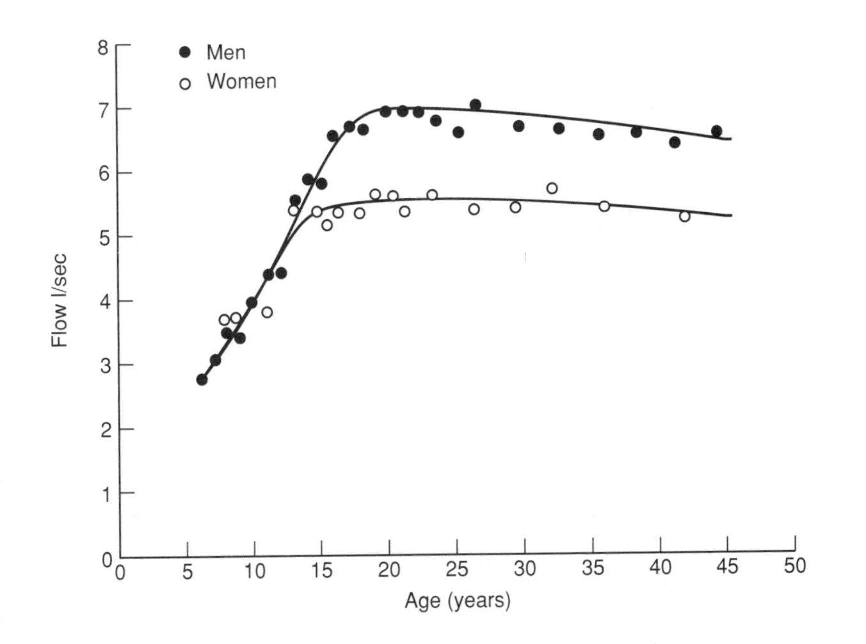

Figure A.2 Distribution of peak flow measurements by age and sex using the Hadorn meter. (After Hadorn, 1942.)

Assess peak flow meter
Personal Best peak flow meter

Electronic meters with facility to measure peak flow
Escort spirometer
Micro Plus spirometer
Micromed pocket spirometer II
VM1 Ventilometer

A.2 Prescribing

The Mini Wright peak flow meter, Vitalograph peak flow meter and Ferraris Pocketpeak peak flow meter are all prescribable in the UK on FP10 in general practice. If a particular type is preferred then this should be specified (or the cheapest will be supplied), and the prescription should also state 'standard' or 'low range' where a choice is possible.

A.3 Technical specifications

A.3.1 Wright peak flow meter (Fig. A.3)

Date: 1959
Weight nett: 1.08 kg (38 oz)
Dimensions: 140 mm × 64 mm
Scale: non-linear
range 60–1000 l/min
minimum scale intervals 5 l/min
Reproducibility: 2 l/min
Calibration accuracy: 10 l/min
Metal casing
Mouthpiece: plastic or disposable cardboard
25.4 mm × 35 mm
Advantages: accurate, simple
Disadvantages: bulky, expensive
Manufacturer: Ferraris Medical Limited

The peak flow meter consists of a pivoted vane whose rotation inside a drum is opposed by air drag and a respiratory spring. A radial inlet, into which a removable mouthpiece can be fitted, leads directly to the vane. A forced expiration through the radial inlet causes the vane to rotate, allowing the breath to escape increasingly from the drum through a peripheral slot. The vane comes to rest at a position which depends on the peak flow which has been attained, and is held at this position by a roller clutch, which can be released by pressing a button adjacent to the radial inlet.

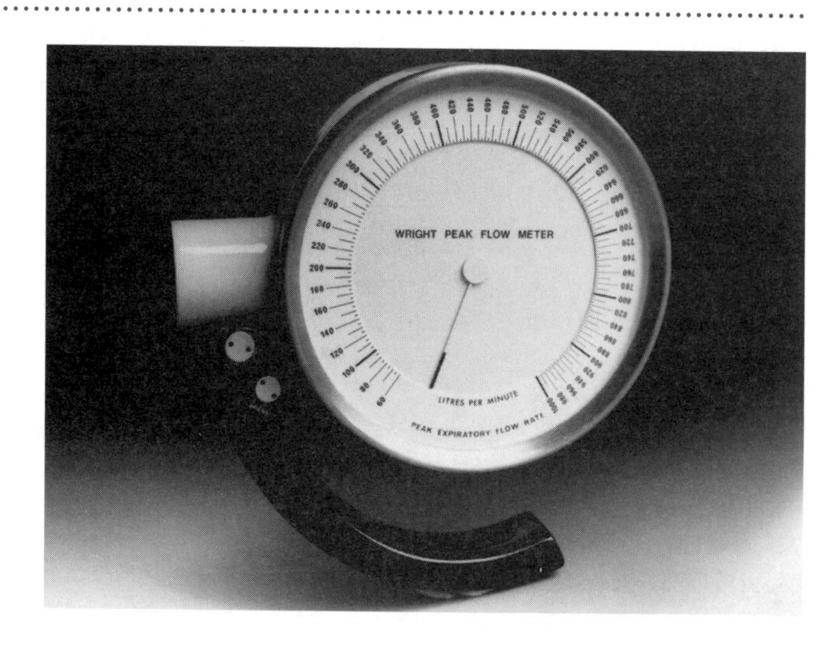

Figure A.3 The Wright peak flow meter.

A.3.2 Mini Wright peak flow meter (Fig. A.4)

Date: 1978 (standard), 1995 (new AFS low range)
Weight nett: 72 g (standard), 52 g (low range)
Dimensions: 200 mm × 44 mm (standard), 147 mm × 30 mm (low range)
Scale: linear
standard range: 60–800 l/min
low reading: 30–400 l/min
Reproducibility: 3%
Plastic casing
Mouthpiece: Standard range, sterilizable plastic
Low range, sterilizable plastic
disposable cardboard for both standard and low ranges
Advantages: portable, cheap, simple
Manufacturer: Clement Clarke International Ltd

The instrument consists of a light plastic cylinder enclosing a spring piston that slides freely on a central rod. The piston drives an indicator along a graduated scale. The piston comes to rest at a level that depends on the maximum flow rate since the air blown into the instrument makes the piston move forward.

Each meter is supplied in a reusable plastic storage container, with instructions for use and a peak flow chart.

Figure A.4 (*a*) The Mini Wright peak flow meter. (*b*) Longitudinal section (reproduced with permission from Wright, 1978).

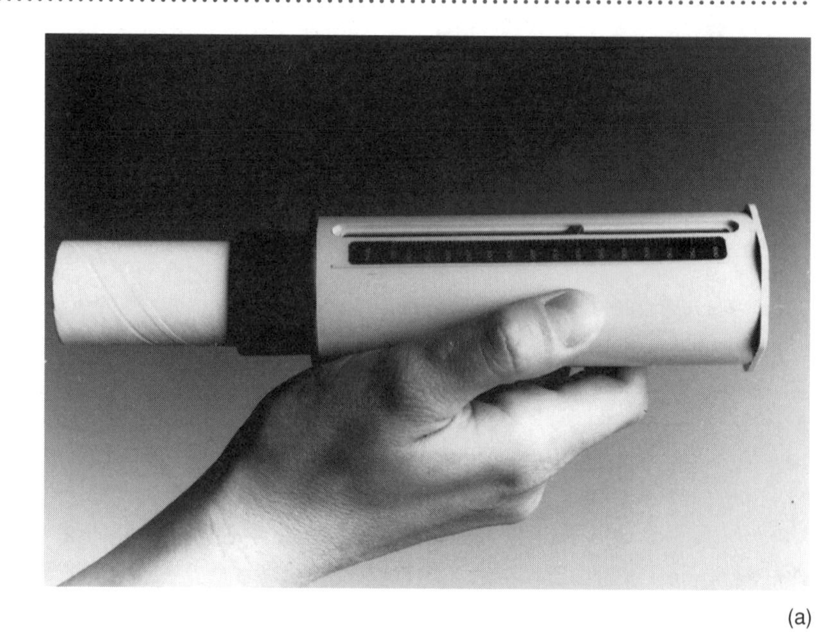

(a)

Marker
Air escapes
Locking nut
Air escapes
Piston
Tension spring
Plastic membrane
Mouthpiece

(b)

A.3.3 Vitalograph peak flow meter (Fig. A.5)

Date: 1976
Weight nett: 74 g
Dimensions: 190 mm × 60 mm × 35 mm
Scale: linear
standard range 50–750 l/min
low range: 25–280 l/min
Reproducibility: standard range 20 l/min
low range 10 l/min
Plastic casing
Mouthpiece: plastic
disposable cardboard
Advantages: portable, cheap, simple
Manufacturer: Vitalograph Ltd

The Vitalograph works on a similar principle to the Mini Wright peak flow meter. Each meter has a plastic carrying pouch, a record chart, and tables of normal values with instructions. A low reading version is available. A colour coded (Asmaplan) version is also available.

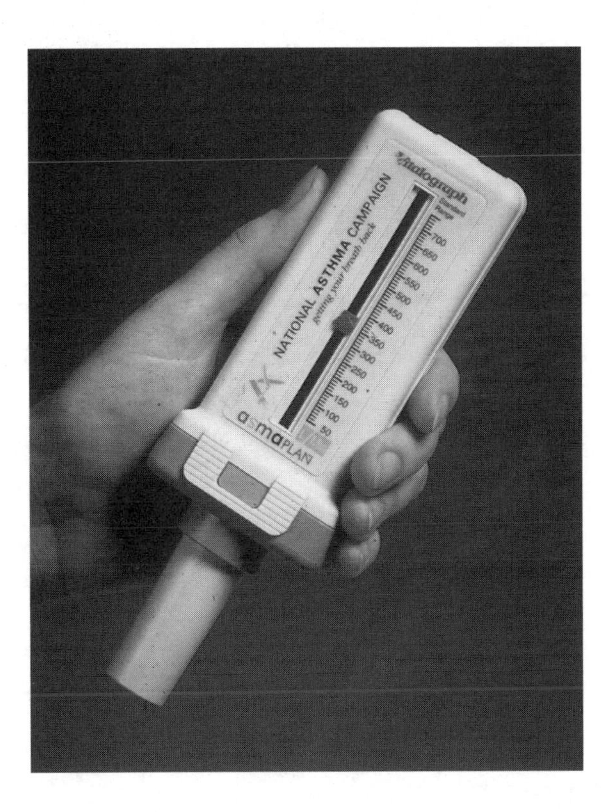

Figure A.5 The Vitalograph peak flow meter.

A.3.4 Ferraris Pocketpeak peak flow meter (Fig. A.6)

Date: 1994
Weight nett: 45 g
Dimensions: 85 mm × 90 mm × 32 mm
Scale: Standard 90–710 l/min
Low range 40–370 l/min
Reproducibility: within 10 l/min
Calibration accuracy: within 10%
Inter-device variability: within 5%
Plastic casing
Mouthpiece: conical plastic mouthpiece for adults or children. Fits either way in standard and low range. Or cylindrical disposable cardboard mouthpiece.
Advantages: compact, simple, portable, lightweight, accurate, cheapest on Drug Tariff, good correlation compared to Wright peak flow meter.
Manufacturer: Ferraris Medical Limited

A cursor registers the movement of a stainless steel vane inside the meter, registering a peak flow reading by moving the cursor up the scale.

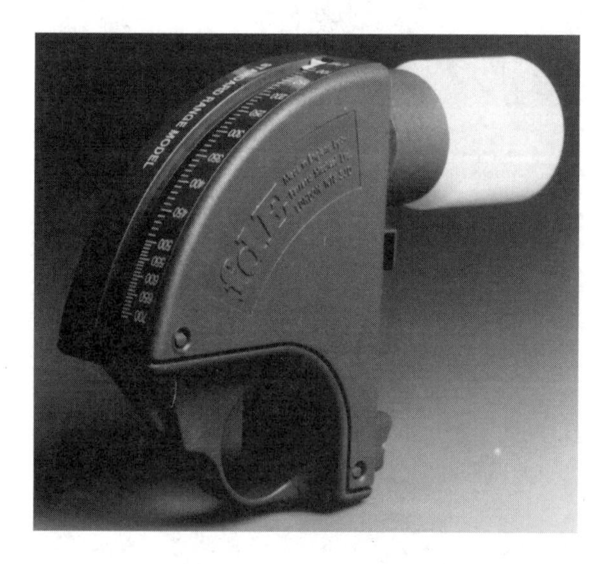

Figure A.6 The Ferraris Pocketpeak peak flow meter.

A.3.5 Assess peak flow meter (Fig. A.7)

Date: 1982
Weight: 73.8 g
Dimensions: 20.3 × 3.3 × 3.2 cm
Scale: Low range 30–390 l/min
Normal range 60–880 l/min
Accuracy: ±5%
Reproducibility: ±1%
Perspex casing
Mouthpiece: plastic
Advantages: portable, cheap, simple
Disadvantages: not available in UK on prescription to date
Manufacturer: Healthscan Products Inc.

This peak flow meter is held vertically.

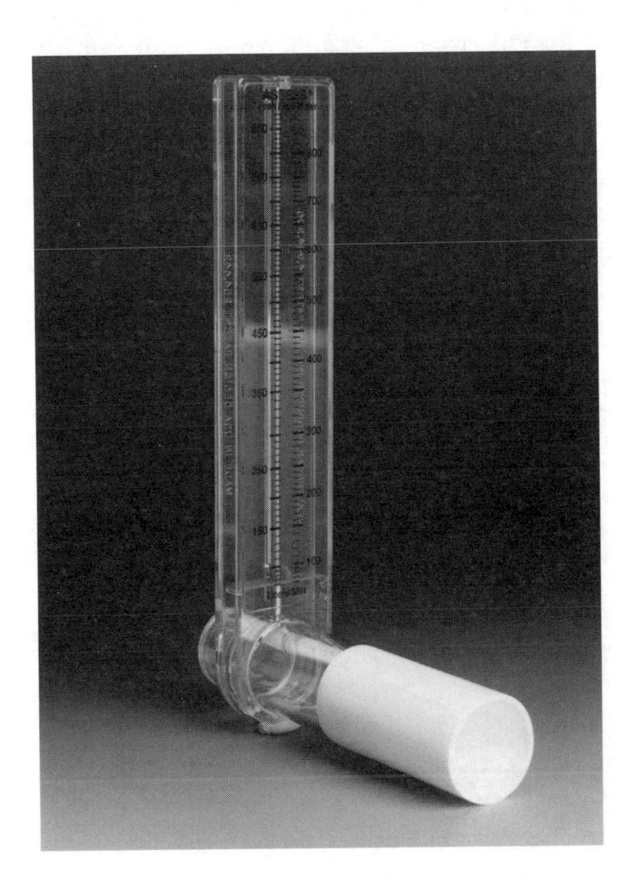

Figure A.7 The Assess peak flow meter.

A.3.6 Personal Best peak flow meter (Fig. A.8)

Date: 1992
Weight: 85.05 g
Dimensions: 165.75 × 50.8 × 20.32 mm

Scale:		standard range	low range
	Wright scale	100–750 l/min	60–450 l/min
	US absolute scale	90–810 l/min	50–390 l/min

Reproducibility: ±1%

	standard range	low range
Calibration accuracy:	10 l/min	5 l/min

Plastic casing
Mouthpiece: oval
Advantages: portable, cheap, simple, absolute and Wright scale available
Disadvantages: not available on prescription in UK to date
Manufacturer: Healthscan Inc. (available in UK from Medix Ltd).

The case forms a fold-out handle and the main body is made of impact-resistant plastic. The cover and handle are of high density polypropylene; the moving parts are stainless steel. The Personal Best is now available in a colour zone version.

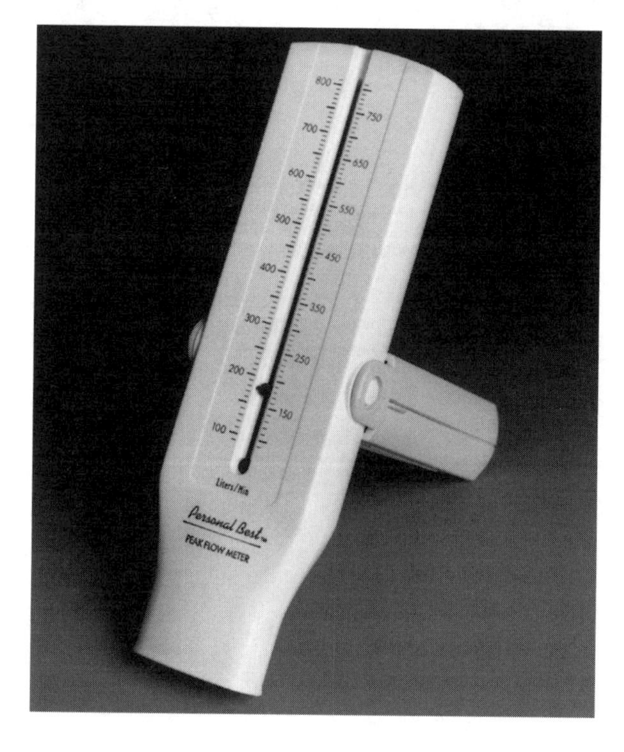

Figure A.8 The Personal Best peak flow meter.

A.3.7 Escort spirometer

Date: 1992
Weight: 450 g
Dimensions: 200 × 130 × 70 mm
Measurements: FEV_1 FVC
FER (forced expiratory ratio)
peak flow
Volume range: 0–12 litres
Flow range: 0–900 l/min
Calibration accuracy: ±5% flow
±3% volume
Power supply: 9 volt rechargeable nickel cadmium.
(Normal operating time when fully
charged 90 minutes)
Advantages: accurate, portable, stores best test result,
drop tested, calibration facility
Disadvantages: expensive
Manufacturer: Vitalograph Ltd.

The Escort has a Fleisch pneumotach flow head with a calibration facility in the form of a hand-held syringe. Calibration, testing and displaying results are performed on three keys. There is a liquid crystal display.

A.3.8 Micro Plus spirometer

Date: 1990
Weight nett: 175 g
Dimensions: 170 mm × 60 mm × 26 mm
Measurements: FEV_1, FVC
FER (forced expiratory ratio)
peak flow
Volume range: 0–9.99 litres
Flow range: 0.1–15.0 l/sec
or 6–900 l/min
Accuracy: 2%
Power supply: single 9 volt PP3 dry cell
Advantages: portable, accurate
Disadvantages: expensive, two separate expiratory manoeuvres required for FVC and peak flow
Manufacturer: Micro Medical

This is a small battery operated spirometer with a liquid crystal display. The instrument comprises a hand-held electronics unit which incorporates a removable digital volume transducer. The transducer consists of a fixed swirl plate which generates rotational flow that drives a low inertia vane. The rotation of this vane interrupts an infra-red beam and produces an electrical pulse train. For a linear unit the volume of air passed through the turbine is proportional to the total number of pulses generated.

The instrument is supplied with a carrying pouch, instruction manual, predicted values, lung function calculator, battery and mouthpieces.

A.3.9 MicroMed II pocket spirometer

Date: 1990
Weight: 450 g
Dimensions: 140 mm × 77 mm × 40 mm
Transducer head: 48 mm × 48 mm × 86 mm
Measurements: FEV_1, FVC, peak flow
percentage of predicted values (ECCS)
for age, height, sex
Volume range: 0.1–10 litres
Scale: 0.1–12 l/sec
Accuracy: 2%
Power supply: single 9 volt PP3 dry cell
Advantages: portable, accurate
Disadvantages: expensive
Manufacturer: Micro Medical

This is a small battery operated spirometer. It has a liquid crystal type display and is simple to use. It has the added facility of displaying percentages of predicted value for age, height and sex which may be keyed in either prior to or after testing.

A.3.10 VM1 Ventilometer

Date: 1990
Weight: 420 g (inc. mouthpiece and battery)
Dimensions: 210 mm × 120 mm × 70 mm
Scale: peak flow 0–800 l/min
FEV$_1$ 0–19.99 litres
FVC 0–10.99 litres
Accuracy: ±5%
Plastic casing
Mouthpiece: plastic or disposable
Advantages: portable, accurate
Disadvantages: expensive
Manufacturer: Clement Clarke International Ltd

A.3.11 VMX Mini-Log (Fig. A.9)

Date: 1992
Weight: 420 g (inc. mouthpiece and battery)
Dimensions: 210 mm × 120 mm × 70 mm
Scale: peak flow 0–800 l/min
Memory: stores over 3 months of peak flow and patient input data
Accuracy: ±5%
Plastic casing
Mouthpiece: plastic or disposable
Advantages: portable, accurate
Disadvantages: expensive
Manufacturer: Clement Clarke International Ltd

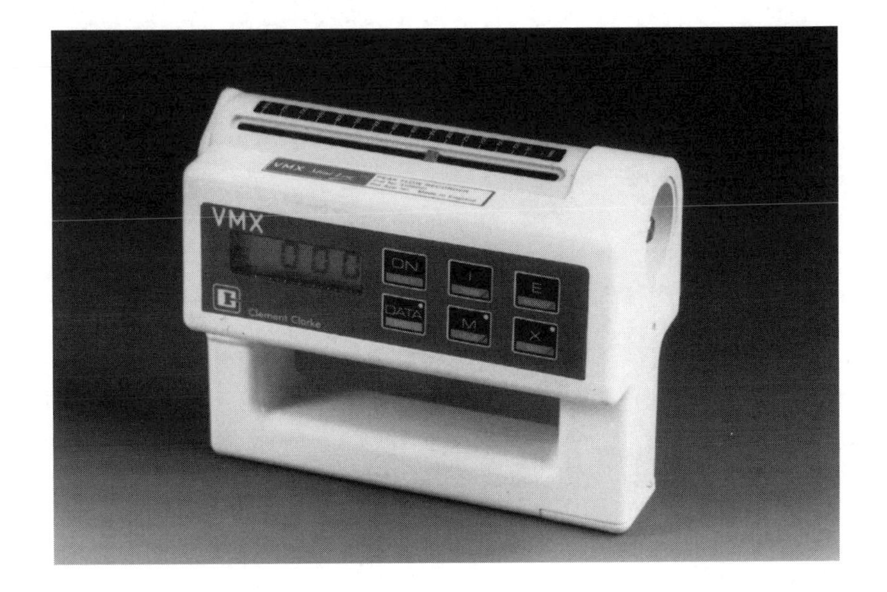

Figure A.9 The VMX Mini-Log.

A.4 Maintenance

With care the mini peak flow meters should last 3 years or more. Peak flow meters should be cleaned regularly. The manufacturers' recommendations vary for different models, but every two weeks would seem reasonable.

Wright peak flow meter This contains a coarse mesh wire gauze filter to prevent mucus entering the body of the machine. The gauze can be unclipped and should be removed regularly and cleaned. In addition to the danger of infection failure to clean the gauze may lead to it clogging up and can falsify readings taken on the meter.

The gauze can be cleaned in a glutaraldehyde or chlorhexidine solution. The meter should be decontaminated by ethylene oxide, formaldehyde or by low temperature steam. It should not be autoclaved.

Mini Wright peak flow meter This can be washed in a mild detergent but not with antiseptic solution. For cleaning inside the meter the locking nut can be unscrewed, but the spring and piston should be left in place (Fig. A.4). The spring does not need lubrication. The danger of cross-infection is probably small because air is not inspired through the meter.

Peak flows result in moist, warm air being blown into the machine which can become a good site for fungal growth if not cleaned. An important component in this respect is the light plastic membrane in the Mini Wright peak flow meter positioned between the mouthpiece and piston diaphragm. A study of mini peak flow meters on loan from an outpatient department showed a filamentous fungus or yeast growing on all the membranes. In heavily contaminated membranes fungal growth was found on both sides and therefore close to the mouth (Fig. A.10).

Vitalograph peak flow meter The meter can be washed in mild detergent and sterilized in Milton crystals or liquid.

Ferraris Pocketpeak peak flow meter This can be cleaned with a mild detergent in warm water (not above 74 °C) and rinsed thoroughly, or even put in a dishwasher. It can be sterilized in low temperature steam (i.e. less than 74 °C).

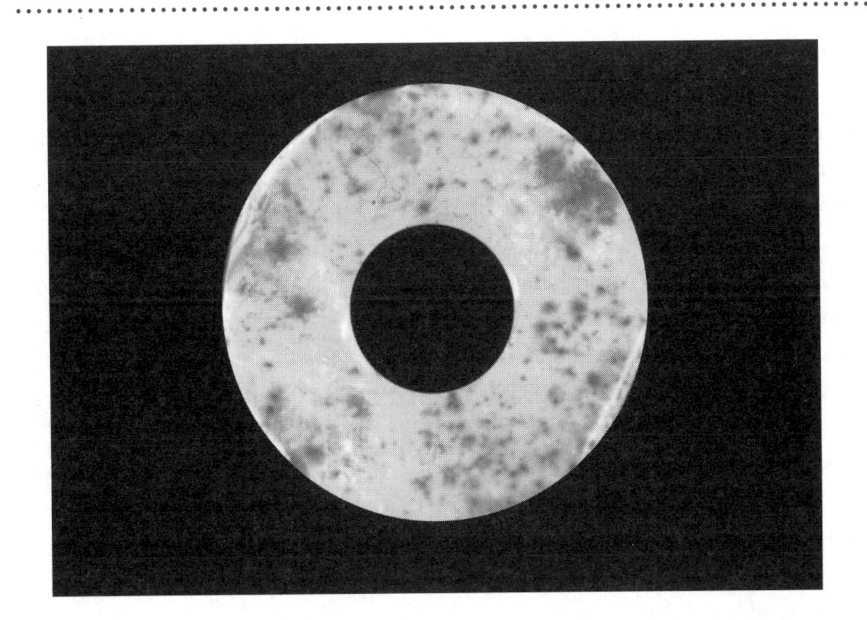

Figure A.10 Fungal growth on a membrane from a Mini Wright peak flow meter. (Reproduced with permission from Ayres *et al.*, 1989.)

In all cases the plastic mouthpieces should be washed and sterilized separately.

Assess No details available.

Personal Best The meter should be washed in soapy water and can be put in a dishwasher.

Escort spirometer The flow head can be dismantled for sterilization or autoclaving. The battery needs to be recharged.

Micro Medical Plus spirometer and pocket spirometer II The manufacturers claim that they are 'virtually maintenance free'. The removable transducer may be immersed in warm soapy water for routine cleaning, or cold sterilizing solutions, although alcohol- and chlorine-based solutions should be avoided.

VMX Mini-Log The Mini Wright component may be cleaned in a mild detergent solution by detaching it from the VMX body. The mouthpiece moulding must not be detached from the VMX, but the detachable mouthpiece may be sterilized in an antiseptic solution.

The 'Intelligent' peak flow meter Peak flow measurements are dependent on the patient's reading of the value and transcription on to a chart or diary card with, at best, an approximation of the exact time of measurement. The 'intelligent' peak flow meter/spirometer manages to avoid these problems by automatically recording the result of a peak flow manoeuvre and the time at which it was performed. The data are recorded on a chip and subsequently downloaded for analysis. The problems of data transfer are thus also avoided and peak flow readings can be made at night if the patient is woken from sleep and without needing to turn on a light. Two versions are now available and are being used both for drug trials and in research, particularly in the diagnosis of occupational asthma, where large quantities of data are required. The two versions are the Diary Card produced by Micro Medical and that produced by Precision Medical. Both can store multiple data points to allow a reasonable length of monitoring time.

A.5 Calibration

The first Wright peak flow meter was developed primarily as an empirical device but a scale was soon employed, calibration having been made in two ways with steady and transient flows:

1. **Steady flow calibration**. The peak flow meter has a low resistance and so can be calibrated by blowing air through it at steady flow rates in series with a rotameter.
2. **Transient flow calibration.** Transient flows can be measured with a pneumotachograph, but this cannot be connected in series with a peak flow meter. Subjects blow alternately into the two instruments, the advantage of which is that the calibration is on human rather than artificially generated peak flows. The peak flows measured on the same person by the two instruments are slightly different but efforts are made to reduce this error by taking a large number of readings.

The original correlation curves between pneumotachograph and peak flow meter were quite consistent and a peak flow meter was

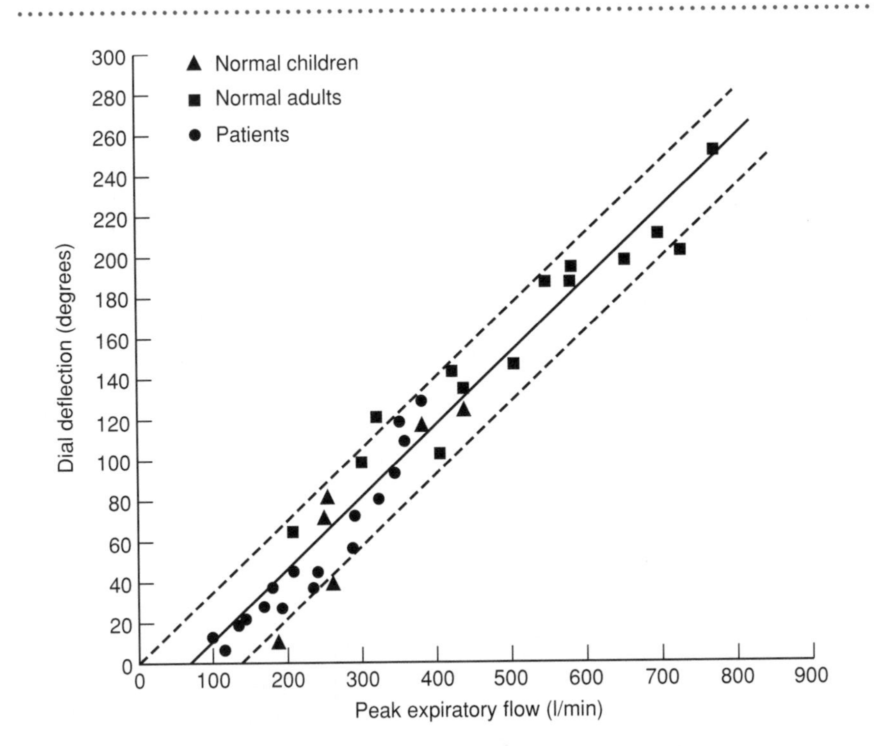

Figure A.11 Calibration curve of Wright peak flow meter readings against pneumotachograph. (Redrawn from Wright and McKerrow, 1959 and reproduced with permission.)

produced with a dial graduated in litres/minute rather than degrees (Fig. A.11).

A.6 Peak flow at altitude

The Mini Wright peak flow meter is sensitive to air density and under-reads at altitude, although true peak flow increases because of the low air density. Evidence suggests that measurements of peak flow at altitude on this instrument should be corrected by the addition of 6.6% of the peak flow value for every 100 mmHg decrease in barometric pressure. Peak flow may be further reduced if the subject is suffering from mild pulmonary oedema caused by acute mountain sickness.

A.7 Variability between meters

Mini Wright peak flow meter and Wright peak flow meter The Mini Wright peak flow meter reads about 3% or 38 l/

min higher than the Wright peak flow meter through most of its range, with some variability between instruments.

Standard and low range Mini Wright peak flow meters The low range recordings tend to be lower than on standard range instruments, especially in the upper part of their range.

Vitalograph peak flow meter and Wright peak flow meter The mean difference between these instruments is 0.57 l/min, mean variation between patients being 5.8%. These variations are unimportant in clinical practice (Fig. A.12).

Mini Wright peak flow meter and turbine spirometer Ninety-eight per cent of readings on the Mini Wright peak flow meter were higher than the corresponding readings on the turbine spirometer, the differences in these readings being wide, ranging from 1 to 173 l/min. In practice this means that these two meters are not interchangeable and again emphasizes the importance of ensuring that the patient measures peak flow from the same instrument (Fig. A.13).

Mini Wright peak flow meter and Ferraris Pocketpeak peak flow meter Preliminary work suggests that the intra-subject

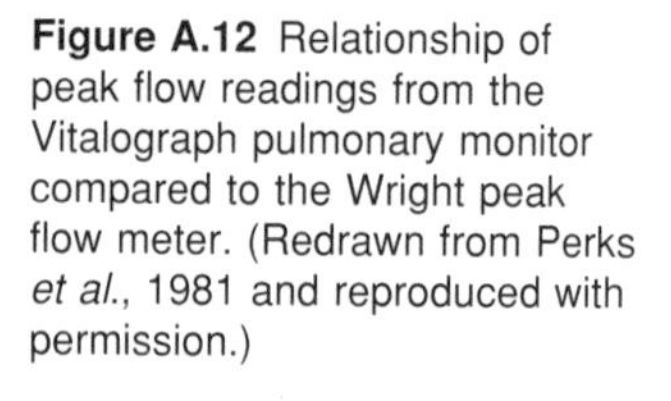

Figure A.12 Relationship of peak flow readings from the Vitalograph pulmonary monitor compared to the Wright peak flow meter. (Redrawn from Perks *et al.*, 1981 and reproduced with permission.)

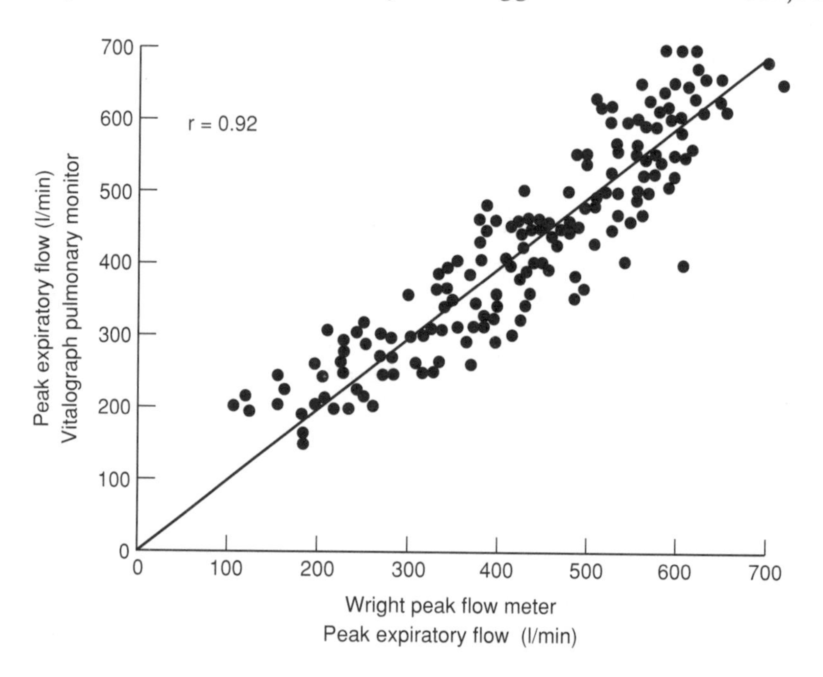

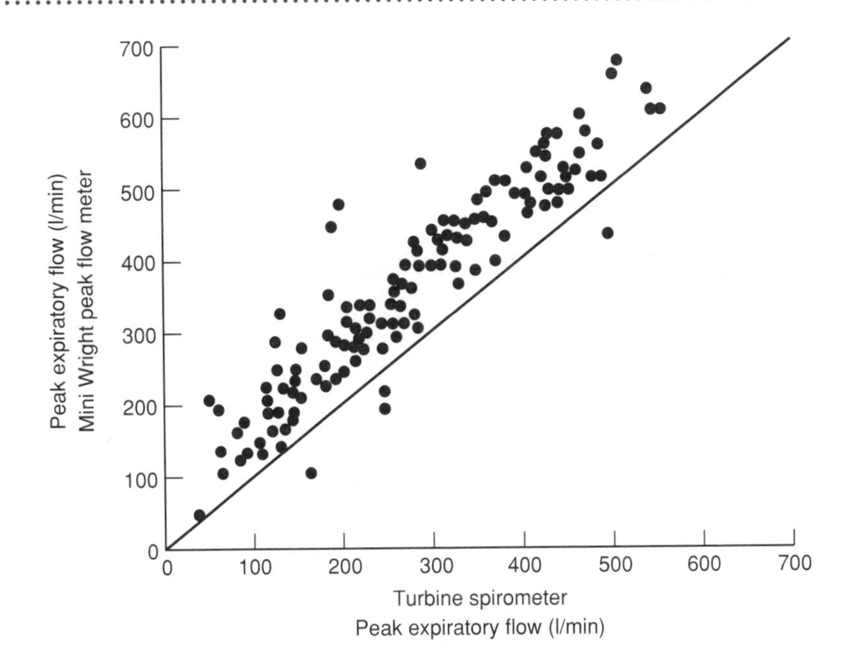

Figure A.13 Comparison of peak flow readings by turbine spirometer to the Mini Wright peak flow meter. (Redrawn from Jones and Mullee, 1990 and reproduced with permission.)

variability in repeated measures of peak flow was much wider with the pocket version. We believe that the pocket meter needs further evaluation.

With normal use the calibration of the meters should stay constant for several years. The Wright peak flow meter and turbine spirometers can be returned to the manufacturer for recalibration if necessary.

On present evidence there is not much to choose between the mini peak flow meters. The Wright peak flow meter and turbine spirometer are largely excluded for day-to-day monitoring purposes on the grounds of expense, although the turbine spirometer is probably the most accurate available.

More recently a study comparing peak flow meter readings against an absolute standard of peak flow generated by computer-driven pumps has shown some remarkable discrepancies between different brands of peak flow meter, although meters of the same type performed with a high degree of repeatability, which is the more important factor in the clinical situation.

Significant over-reading of peak flow occurred in the range 300–500 l/min with overestimating of up to 80 l/min in Ferraris Pocketpeak and Mini Wright peak flow meters (Figs A.14, A.15). Discrepancies between true and recorded peak flow were less marked for the Vitalograph peak flow meter and original Wright peak flow meter. The

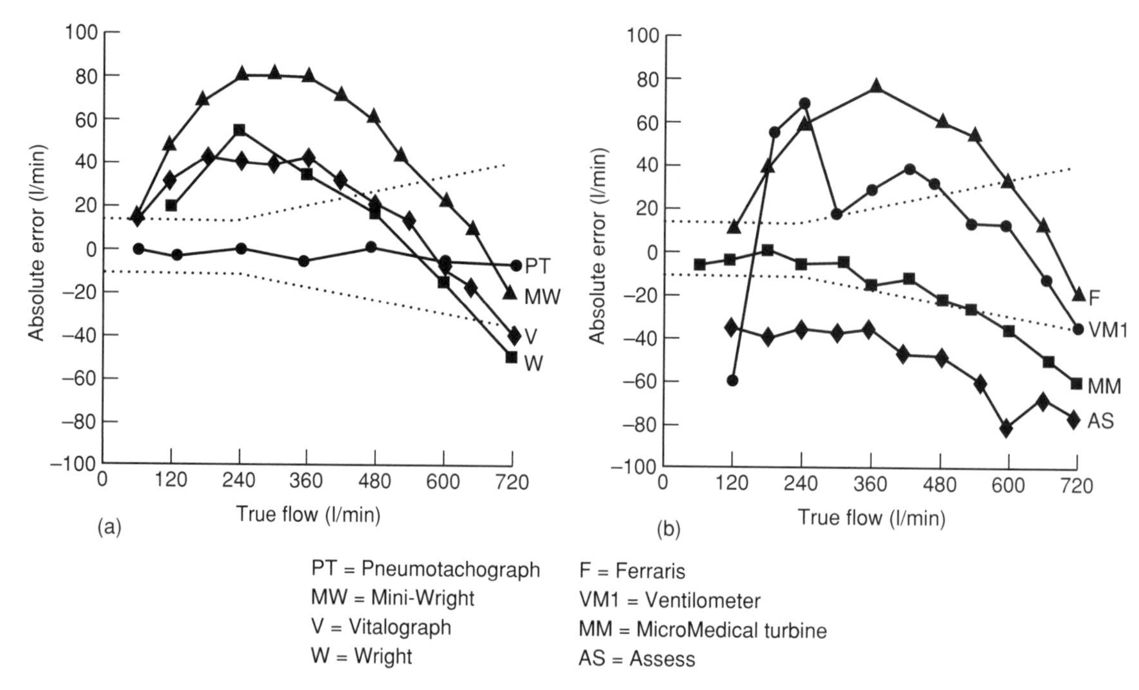

Figure A.14 Absolute error plots for the range of peak flow meters. The dotted lines indicate the American Thoracic Society guidelines for accuracy. PT, pneumotachograph; MW, Mini Wright; V, Vitalograph; W, Wright; F, Ferraris; VM1, Ventilometer; MM, MicroMedical Turbine; AS, Assess. (Redrawn from Miller *et al.*, 1992 and reproduced with permission.)

Figure A.15 Comparison of peak flow readings on a Ferraris Pocketpeak peak flow meter compared to a Wright peak flow meter. (Redrawn from Nolan *et al.*, 1992 and reproduced with permission.)

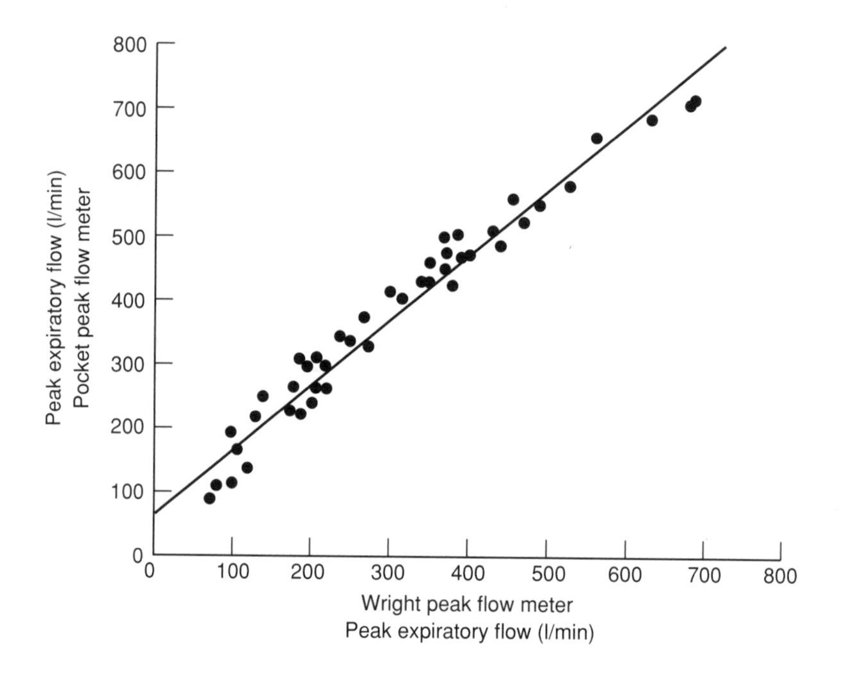

MicroMed turbine spirometer was accurate up to 400 l/min but above this level tended increasingly to under-read.

This variation in performance has potential implications for the use of peak flow in guiding management in asthma (see Chapter 5). Correction to true flow results in significant changes in setting appropriate guideline peak flow targets and also the frequency with which the patients dip below such targets and thus need to alter treatment. Although the planned introduction of corrected scales for the meters will improve this aspect of peak flow monitoring, these findings should in no way deter the use of the current peak flow meters in the management of asthma. The overall information about peak flow and its pattern of variability will remain the same.

References and further reading

Ayres, J.G., Whitehead, J., Boldy, D.A.R. and Sykes, A. (1989) Fungal contamination of mini peak flow meters. *Respir. Med.*, **83**, 503–4.

Bodani, H., Pearson, J. and Levy, M. (1992) Comparison of the Mini-Wright and the Wright Pocket peak flow meters. *Thorax*, 4, 230.

Donald, K.W. (1953) The definition and measurement of respiratory function. *Br. Med. J.*, i, 415–22.

Fisher, J. and Shaw, A. (1980) Calibration of some Wright peak flow meters. *Br. J. Anaesth.*, **53**, 461–4.

Hadorn, W. (1942) Ein neues pneumaneter zur Bestimmung der Exspirationstosse (maximal Ausatmungsstranstarte). *Schweiz. Med. Wochenschr.*, **23**, 946–50.

Haydn, S.P., Chapman, T.T. and Hughes, D.T.D. (1976) Pulmonary monitor for assessment of airways obstruction. *Lancet*, **ii**, 1225.

Jones, K.P. and Mullee, M.A. (1990) Measuring peak expiratory flow in general practice: comparison of Mini-Wright peak flow meter and turbine spirometer. *Br. Med. J.*, **300**, 1629–31.

Miller, M.R., Dickinson, S.A. and Hitchings, D.S. (1992) The accuracy of portable peak flow meters. *Thorax*, 47, 904–9.

Nolan, K.M., Dornelly, S.M., Hughes, D.T. and Strunin, L. (1992) Evaluation of the pocket peak flow meter for the measurement of peak expiratory flow rate (PEFR), and forced expiratory volume in the first second (FEV_1). *Respir. Med.*, **86**, 525–6.

Oldham, H.G., Bevan, M.M. and McDermott, M. (1979) Comparison of the new miniature Wright peak flow meter with the standard Wright peak flow meter. *Thorax*, 34, 807–9.

Perks, W.H., Cole, M., Steventon, R.D., Tams, I.P. and Prowse, K. (1981) An evaluation of the Vitalograph pulmonary monitor. *Br. J. Dis. Chest*, 75, 161–4.

Perks, W.H., Tams, I.P., Thompson, D.A. and Prowse, K. (1979) An evaluation of the Mini-Wright peak flow meter. *Thorax*, 34, 78–81.

Shephard, R.J. (1955) Pneumotachographic measurement of breathing capacity. *Thorax*, **10**, 258–60.

Usherwood, T.P. and Barber, J.H. Discrepancy between standard and low range Mini Wright peak flow meters. *Br. Med. J.*, **292**, 523–4.

Wright, B.M. (1978) A miniature Wright peak flow meter. *Br. Med. J.*, ii, 1627–8.

Wright, B.M. (1981) Calibration of peak flow meters. *Br. J. Anaesth.*, 53, 777.

Wright, B.M. and McKerrow, C.B. (1959) Maximum forced expiratory flow rate as a measure of ventilatory capacity. *Br. Med. J.*, ii, 1041–7.

Predicted values (ECCS) for caucasian males and females

Appendix B

Predicted values for males, 25 years

Ht (cm)	*150*	*155*	*160*	*165*	*170*	*175*	*180*	*185*	*190*	*195*
FEV_1 (l)	3.24	3.45	3.67	3.88	4.10	4.31	4.53	4.74	4.96	5.17
FVC (l)	3.65	3.94	4.23	4.51	4.80	5.09	5.38	5.67	5.95	6.24
FEV_1 (%)	89.0	88.0	87.0	86.0	85.0	85.0	84.0	84.0	83.0	83.0
PEF (l/m)	496.0	515.0	533.0	552.0	570.0	588.0	607.0	625.0	644.0	662.0

Predicted values for males, 30 years

Ht (cm)	*150*	*155*	*160*	*165*	*170*	*175*	*180*	*185*	*190*	*195*
FEV_1 (l)	3.09	3.30	3.52	3.73	3.95	4.16	4.38	4.59	4.81	5.03
FVC (l)	3.52	3.81	4.10	4.38	4.67	4.96	5.25	5.54	5.82	6.11
FEV_1 (%)	88.0	87.0	86.0	85.0	85.0	84.0	83.0	83.0	83.0	82.0
PEF (l/m)	483.0	502.0	520.0	539.0	557.0	575.0	594.0	612.0	631.0	649.0

Predicted values for males, 35 years

Ht (cm)	*150*	*155*	*160*	*165*	*170*	*175*	*180*	*185*	*190*	*195*
FEV_1 (l)	2.95	3.16	3.37	3.59	3.80	4.02	4.23	4.45	4.67	4.88
FVC (l)	3.39	3.68	3.97	4.25	4.54	4.83	5.12	5.41	5.69	5.98
FEV_1 (%)	87.0	86.0	85.0	84.0	84.0	83.0	83.0	82.0	82.0	82.0
PEF (l/m)	470.0	489.0	507.0	526.0	544.0	562.0	581.0	599.0	618.0	636.0

Predicted values for males, 40 years

Ht (cm)	*150*	*155*	*160*	*165*	*170*	*175*	*180*	*185*	*190*	*195*
FEV_1 (l)	2.80	3.02	3.23	3.45	3.66	3.87	4.09	4.30	4.52	4.74
FVC (l)	3.26	3.55	3.84	4.12	4.41	4.70	4.99	5.28	5.56	5.85
FEV_1 (%)	86.0	85.0	84.0	84.0	83.0	82.0	82.0	82.0	81.0	81.0
PEF (l/m)	457.0	476.0	494.0	512.0	531.0	549.0	568.0	586.0	605.0	623.0

Predicted values for males, 45 years

Ht (cm)	*150*	*155*	*160*	*165*	*170*	*175*	*180*	*185*	*190*	*195*
FEV_1 (l)	2.66	2.87	3.09	3.30	3.52	3.73	3.95	4.16	4.38	4.59
FVC (l)	3.13	3.42	3.71	3.99	4.28	4.57	4.86	5.15	5.43	5.72
FEV_1 (%)	85.0	84.0	83.0	83.0	82.0	82.0	81.0	81.0	81.0	80.0
PEF (l/m)	444.0	463.0	481.0	499.0	518.0	536.0	555.0	573.0	592.0	610.0

Predicted values for males, 50 years

Ht (cm)	*150*	*155*	*160*	*165*	*170*	*175*	*180*	*185*	*190*	*195*
FEV_1 (l)	2.51	2.73	2.94	3.16	3.37	3.59	3.80	4.02	4.23	4.45
FVC (l)	3.00	3.29	3.58	3.86	4.15	4.44	4.73	5.02	5.30	5.59
FEV_1 (%)	84.0	83.0	82.0	82.0	81.0	81.0	80.0	80.0	80.0	79.0
PEF (l/m)	431.0	450.0	468.0	486.0	505.0	523.0	542.0	560.0	578.0	597.0

Predicted values for males, 55 years

Ht (cm)	*150*	*155*	*160*	*165*	*170*	*175*	*180*	*185*	*190*	*195*
FEV_1 (l)	2.37	2.58	2.79	3.01	3.22	3.44	3.65	3.87	4.09	4.30
FVC (l)	2.87	3.16	3.45	3.73	4.02	4.31	4.60	4.89	5.17	5.46
FEV_1 (%)	82.0	82.0	81.0	81.0	80.0	80.0	79.0	79.0	79.0	79.0
PEF (l/m)	418.0	436.0	455.0	473.0	492.0	510.0	529.0	547.0	565.0	584.0

Predicted values for males, 60 years

Ht (cm)	*150*	*155*	*160*	*165*	*170*	*175*	*180*	*185*	*190*	*195*
FEV_1 (l)	2.22	2.44	2.65	2.87	3.08	3.29	3.51	3.72	3.94	4.16
FVC (l)	2.74	3.03	3.32	3.60	3.89	4.18	4.47	4.76	5.04	5.33
FEV_1 (%)	81.0	80.0	80.0	79.0	79.0	79.0	79.0	78.0	78.0	78.0
PEF (l/m)	405.0	423.0	442.0	460.0	479.0	497.0	516.0	534.0	552.0	571.0

Predicted values for males, 65 years

Ht (cm)	*150*	*155*	*160*	*165*	*170*	*175*	*180*	*185*	*190*	*195*
FEV_1 (l)	2.08	2.29	2.51	2.72	2.94	3.15	3.37	3.58	3.80	4.01
FVC (l)	2.61	2.90	3.19	3.47	3.76	4.05	4.34	4.63	4.91	5.20
FEV_1 (%)	80.0	79.0	79.0	78.0	78.0	78.0	78.0	77.0	77.0	77.0
PEF (l/m)	392.0	410.0	429.0	447.0	466.0	484.0	502.0	521.0	539.0	558.0

Predicted values for males, 70 years

Ht (cm)	*150*	*155*	*160*	*165*	*170*	*175*	*180*	*185*	*190*	*195*
FEV_1 (l)	1.93	2.14	2.36	2.57	2.79	3.00	3.22	3.43	3.65	3.87
FVC (l)	2.48	2.77	3.06	3.34	3.63	3.92	4.21	4.50	4.78	5.07
FEV_1 (%)	78.0	77.0	77.0	77.0	77.0	77.0	77.0	76.0	76.0	76.0
PEF (l/m)	379.0	397.0	416.0	434.0	453.0	471.0	489.0	508.0	526.0	545.0

Predicted values for males, 75 years

Ht (cm)	150	155	160	165	170	175	180	185	190	195
FEV_1 (l)	1.79	2.00	2.21	2.43	2.64	2.86	3.07	3.29	3.51	3.72
FVC (l)	2.35	2.64	2.93	3.21	3.50	3.79	4.08	4.37	4.65	4.94
FEV_1 (%)	76.0	76.0	76.0	76.0	76.0	75.0	75.0	75.0	75.0	75.0
PEF (l/m)	366.0	384.0	403.0	421.0	440.0	458.0	476.0	495.0	513.0	532.0

Predicted values for males, 80 years

Ht (cm)	150	155	160	165	170	175	180	185	190	195
FEV_1 (l)	1.64	1.86	2.07	2.29	2.50	2.71	2.93	3.14	3.36	3.58
FVC (l)	2.22	2.51	2.80	3.08	3.37	3.66	3.95	4.24	4.52	4.81
FEV_1 (%)	74.0	74.0	74.0	74.0	74.0	74.0	74.0	74.0	74.0	74.0
PEF (l/m)	353.0	371.0	390.0	408.0	426.0	445.0	463.0	482.0	500.0	519.0

Predicted values for females, 25 years

Ht (cm)	140	145	150	155	160	165	170	175	180	185
FEV_1 (l)	2.30	2.50	2.70	2.90	3.10	3.29	3.49	3.69	3.88	4.08
FVC (l)	2.66	2.88	3.10	3.33	3.55	3.77	3.99	4.21	4.43	4.66
FEV_1 (%)	87.0	87.0	87.0	87.0	87.0	87.0	87.0	88.0	88.0	88.0
PEF (l/m)	350.0	367.0	383.0	400.0	416.0	433.0	449.0	466.0	482.0	499.0

Predicted values for females, 30 years

Ht (cm)	140	145	150	155	160	165	170	175	180	185
FEV_1 (l)	2.18	2.38	2.56	2.77	2.97	3.17	3.37	3.56	3.76	3.96
FVC (l)	2.53	2.75	2.98	3.20	3.42	3.64	3.86	4.08	4.30	4.53
FEV_1 (%)	86.0	86.0	87.0	87.0	87.0	87.0	87.0	87.0	87.0	87.0
PEF (l/m)	341.0	358.0	374.0	391.0	407.0	424.0	440.0	457.0	473.0	490.0

Predicted values for females, 35 years

Ht (cm)	140	145	150	155	160	165	170	175	180	185
FEV_1 (l)	2.05	2.25	2.45	2.65	2.85	3.04	3.24	3.44	3.63	3.83
FVC (l)	2.40	2.82	2.84	3.07	3.29	3.51	3.73	3.95	4.17	4.40
FEV_1 (%)	86.0	86.0	86.0	86.0	87.0	87.0	87.0	87.0	87.0	87.0
PEF (l/m)	332.0	349.0	365.0	382.0	398.0	415.0	431.0	448.0	464.0	481.0

Predicted values for females, 40 years

Ht (cm)	140	145	150	155	160	165	170	175	180	185
FEV_1 (l)	1.93	2.13	2.33	2.51	2.72	2.92	3.12	3.31	3.51	3.71
FVC (l)	2.27	2.49	2.71	2.94	3.16	3.38	3.60	3.82	4.04	4.27
FEV_1 (%)	85.0	85.0	86.0	86.0	86.0	86.0	87.0	87.0	87.0	87.0
PEF (l/m)	323.0	346.0	356.0	373.0	389.0	406.0	422.0	439.0	455.0	472.0

Predicted values for females, 45 years

Ht (cm)	140	145	150	155	160	165	170	175	180	185
FEV_1 (l)	1.60	2.00	2.20	2.40	2.60	2.79	2.99	3.19	3.38	3.58
FVC (l)	2.14	2.36	2.58	2.81	3.03	3.25	3.47	3.69	3.91	4.14
FEV_1 (%)	84.0	85.0	85.0	85.0	86.0	86.0	86.0	86.0	86.0	87.0
PEF (l/m)	314.0	331.0	347.0	364.0	380.0	397.0	413.0	430.0	446.0	463.0

Predicted values for females, 50 years

Ht (cm)	140	145	150	155	160	165	170	175	180	185
FEV_1 (l)	1.68	1.88	2.08	2.27	2.47	2.67	2.87	3.06	3.26	3.46
FVC (l)	2.01	2.23	2.46	2.68	2.90	3.12	3.34	3.56	3.78	4.01
FEV_1 (%)	83.0	84.0	85.0	85.0	85.0	86.0	86.0	86.0	86.0	86.0
PEF (l/m)	305.0	322.0	338.0	355.0	371.0	388.0	404.0	421.0	437.0	454.0

Predicted values for females, 55 years

Ht (cm)	140	145	150	155	160	165	170	175	180	185
FEV_1 (l)	1.55	1.75	1.95	2.15	2.35	2.54	2.74	2.94	3.13	3.33
FVC (l)	1.88	2.10	2.33	2.55	2.77	2.99	3.21	3.43	3.65	3.88
FEV_1 (%)	83.0	83.0	84.0	84.0	85.0	85.0	85.0	86.0	86.0	86.0
PEF (l/m)	296.0	313.0	329.0	346.0	362.0	379.0	395.0	412.0	428.0	445.0

Predicted values for females, 60 years

Ht (cm)	140	145	150	155	160	165	170	175	180	185
FEV_1 (l)	1.43	1.63	1.83	2.02	2.22	2.42	2.62	2.81	3.01	3.21
FVC (l)	1.75	1.97	2.19	2.42	2.64	2.86	3.08	3.30	3.52	3.75
FEV_1 (%)	82.0	82.0	83.0	84.0	84.0	85.0	85.0	85.0	85.0	86.0
PEF (l/m)	287.0	304.0	320.0	337.0	353.0	370.0	386.0	403.0	419.0	436.0

Predicted values for females, 65 years

Ht (cm)	140	145	150	155	160	165	170	175	180	185
FEV_1 (l)	1.30	1.50	1.70	1.90	2.10	2.29	2.49	2.69	2.88	3.08
FVC (l)	1.62	1.84	2.07	2.29	2.51	2.75	2.95	3.17	3.39	3.62
FEV_1 (%)	80.0	82.0	82.0	83.0	84.0	84.0	84.0	85.0	85.0	85.0
PEF (l/m)	276.0	295.0	311.0	328.0	344.0	361.0	377.0	394.0	410.0	427.0

Predicted values for females, 70 years

Ht (cm)	140	145	150	155	160	165	170	175	180	185
FEV_1 (l)	1.18	1.38	1.58	1.77	1.97	2.17	2.37	2.56	2.76	2.96
FVC (l)	1.49	1.71	1.93	2.16	2.38	2.60	2.82	3.04	3.26	3.49
FEV_1 (%)	79.0	80.0	81.0	82.0	83.0	83.0	84.0	84.0	85.0	85.0
PEF (l/m)	269.0	286.0	302.0	319.0	335.0	352.0	368.0	385.0	401.0	418.0

Predicted values for females, 75 years

Ht (cm)	140	145	150	155	160	165	170	175	180	185
FEV_1 (l)	1.05	1.25	1.45	1.65	1.85	2.04	2.24	2.44	2.63	2.83
FVC (l)	1.36	1.58	1.81	2.03	2.25	2.47	2.69	2.91	3.13	3.36
FEV_1 (%)	77.0	79.0	80.0	81.0	82.0	83.0	83.0	84.0	84.0	84.0
PEF (l/m)	260.0	277.0	293.0	310.0	326.0	343.0	359.0	376.0	392.0	409.0

Predicted values for females, 80 years

Ht (cm)	140	145	150	155	160	165	170	175	180	185
FEV_1 (l)	0.93	1.13	1.33	1.52	1.72	1.92	2.12	2.31	2.51	2.71
FVC (l)	1.23	1.45	1.67	1.90	2.12	2.34	2.56	2.78	3.00	3.23
FEV_1 (%)	75.0	78.0	79.0	80.0	81.0	82.0	83.0	83.0	84.0	84.0
PEF (l/m)	251.0	266.0	264.0	301.0	317.0	334.0	350.0	367.0	383.0	400.0

Reference

Quanjer, P. (1983) Standardized lung function testing. *Bull. Eur. Physiopathol. Respir.*, 19 (Suppl. 5), 1–9.

Appendix C Diurnal variation calculator

Maximum daily peak flow (columns) / Minimum daily peak flow (rows)

Min \ Max	100	125	150	175	200	225	250	275	300	325	350	375	400	425	450	475	500	525	550	575	600
100	0	20	33	43	50	55	60	63	67	69	71	73	75	76	78	79	80	81	82	83	83
125		0	17	29	37	44	50	55	58	61	64	67	69	71	72	74	75	76	77	78	79
150			0	14	25	33	40	45	50	54	57	60	62	65	67	68	70	71	73	74	78
175				0	12	22	30	36	42	46	50	53	56	59	61	63	65	67	68	70	74
200					0	11	20	27	33	38	43	47	50	53	56	58	60	62	64	65	70
225						0	10	18	25	31	36	40	44	47	50	53	55	57	59	61	65
250							0	9	17	23	29	33	37	41	44	47	50	52	55	57	61
275								0	8	15	21	27	31	35	39	42	45	48	50	52	56
300									0	8	14	20	25	29	33	37	40	43	45	48	50
325										0	7	13	19	23	28	32	35	38	41	43	46
350											0	7	12	18	22	26	30	33	36	39	42
375												0	6	12	17	21	25	29	32	35	37
400													0	6	11	16	20	24	27	30	33
425														0	6	11	15	19	23	26	29
450															0	5	10	14	18	22	25
475																0	5	10	14	17	21
500																	0	5	9	13	17
525																		0	5	9	13
550																			0	4	8
575																				0	4
600																					0

Reading off daily maximum and minimum peak flow gives diurnal variation expressed as a percentage

$$DV\% = \frac{Max - Min}{Max} \%$$

Key
A > 50% diurnal variation
B > 20% but <50% diurnal variation
C < 20% diurnal variation

Cosinor analysis

Appendix D

This technique uses a least squares method to test the goodness of fit of the raw data to a sinusoidal waveform. The function:

$$f(t) = \text{Co} + \text{C.Cos}\left(\frac{2\pi t}{T} + \phi\right) + \varepsilon_t$$

Where t = time, f = biological variable under study, Co = constant term or intercept, C = amplitude, ϕ = phase, T = trial period under study, ε_t = residual error (when mean of $\varepsilon_t = 0$ and standard deviation of ε_t = standard error of estimate) is fitted to the raw data for different trial periods, T. When the subject under study is synchronized to the solar day, T can be assumed to be 24 hours. The rhythm is most conveniently represented as a cosine wave, if the reference point for phase is 0° or 00.00 hours, since $\cos 0° = 1$. A multiple regression programme is used to analyse peak expiratory flow against time, using the equation:

$$\text{PEF} = \text{Co} + a\cos\left(\frac{2\pi t}{24}\right) + b\sin\left(\frac{2\pi t}{24}\right)$$

Zero time is taken as 00.00 hours on the first day of the study. Amplitude, C, and phase are determined from the coefficients a and b of the $\cos(\frac{2\pi t}{24})$ and $\sin(\frac{2\pi t}{24})$ terms in the regression equation. The amplitude, C, in this mathematical model represents half the difference observed between the highest and lowest values in a complete cycle (360° or 24 hours). From a clinical viewpoint, however, the peak to trough measurement of the PEF rhythm – that is, the maximum change in PEF during the 24 hour cycle – is more important (Halberg *et al.*, 1964). Here 'amplitude' is defined as the peak to trough measurement and $= 2C = 2\sqrt{a^2 + b^2}$.

The phase of the rhythm is identified conventionally as the time of the computed acrophase (peak reading in the 24 hour cycle) and is determined by the equation: phase $\phi° = \arctan \frac{(-b)}{a}$.

The amplitude (peak to trough measurement) of each individual subject's PEF rhythm should be expressed as a percentage of each individual subject's mean PEF over the study period to facilitate comparison between normal subjects and asthma patients with widely differing peak flow rates and predicted normal values.

This method is a research tool and probably only applied to a specific subset of asthmatic patients and has no place in the day-to-day management of asthma.

References and further reading

Halberg, F. *et al.* (1964) Compared techniques in the study of biological rhythms. *Ann. N.Y. Acad. Sci.*, **115**, 695–720.

Hetzel, M.R. and Clark, D.H. (1980) Comparison of normal and asthmatic circadian rhythm in peak expiratory flow rate. *Thorax*, **35**, 732–8.

Regression equations for calculation of peak expiratory flow by nationality/race

Appendix E

A = adults C = children M = male F = female
h = height a = age PPF = predicted peak flow

Authors	*Population*	*Regression*	*Comments*
Nairn *et al.* (1961)	Scottish (children)	M&F: PPF = $5.47h - 460$	
Leiner *et al.* (1963)	US, white (adults)	M: PPF = $(3.95 - 0.0151a)h$ F: PPF = $(2.93 - 0.0072a)h$	Includes smokers, also provides nomograms
Ferris *et al.* (1965)	US, white	M: PPF = $4.73h - 2.46a - 200.32$ F: PPF = $2.96h - 1.71a - 39.19$	No disease, standing height
		M: PPF = $3.90h - 3.00a - 49.36$ F: PPF = $2.63h - 1.84a - 14.06$	Smokers, standing height
Pelzer and Thompson (1964)	UK, white (adults)	F: PPF = $3.74 \times h - 2.117 \times a - 113$ M: PPF = $4.054 \times h - 2.824 \times a + 6$	Small numbers, includes smokers
Johannsen and Grasmus (1968)	Bantu (adults)	M: PPF = $37.89 + 2.92h - 4.19a$ F: PPF = $-42.93 + 2.07h$	Age 20–50 yr, includes smokers
Juhl (1970)	Danish (children)	M: PPF = $5.24h - 439$ F: PPF = $4.95h - 407$ PPF = $5.1h - 425$	Age 7–17 yr
Kallqvist *et al.* (1970)	Swedish (adults)	M: PPF = $282.052 + 2.227h - 2.91a$ F: PPF = $211.173 + 1.971h - 2.59a$	Age 45–65, includes smokers
Elabute and Femi-Pearse (1971)	Nigerian (black adults)	M: PPF = $-41.5 + 3.47h - 1.73a$ F: PPF = $673.2 - 1.07h - 3.99a$	Includes smokers, age 20–55 yr

Regression equations for calculation of peak expiratory flow by nationality/race *(Continued)*

Authors	*Population*	*Regression*	*Comments*
Malik *et al.* (1972)	W. Pakistan workers in UK	M: PPF = −16.34 − 3.72*a* + 3.67*h*	Age 25–60 yr Nomograms
Cookson *et al.* (1976)	Rhodesian Africans (adults)	M: PPF = 189.27 + 2.311*h* − 2.197*a* F: PPF = 100.54 + 2.819*h* − 3.730*a*	Age 20 and over, includes smokers
Malik and Banga (1978)	India (Punjab, adults)	M: PPF = 24.37 + 1.18*a* − 0.34a^2 + 2.29*h*	Non-smokers
Lam *et al.* (1982)	Hong Kong Chinese (children and adults)	MA: PPF = 0.86*a* − 0.00047*a*3 + 313.4 FA: PPF = 0.45*a* − 0.00032*a*3 + 259.7	Adults 15–74 yr F 22–74 yr M includes smokers
		MC: PPF = 15.08*a* − 0.0075*a*3 + 75.5 FC: PPF = 19.96*a* − 0.0209*a*3 + 33.8	'Children' = 5–22M 5–15F some smokers
Nunn and Gregg (1989)	UK, white	M: $\log_e \mathrm{PPF} = 0.544 \log_e a - 0.0151a - \frac{74.7}{h(\mathrm{cm})} + 5.48$	Age 55+
		W: $\log_e \mathrm{PPF} = 0.376 \log_e a - 0.012a - \frac{58.8}{h(\mathrm{cm})} + 5.63$	Age 65+, non-smokers

References

Cookson, J.B., Blake, G.T.W. and Favanisi, C. *et al.* (1976) Normal values for ventilatory function in Rhodesian Africans. *Br. J. Dis. Chest*, 70, 107–11.

Elabute, E.A. and Femi-Pearse, D. (1971) Peak flow rate in Nigeria. *Thorax*, 26, 597–601.

Ferris, B.G., Anderson, D.O. and Zickmantel, R. (1965) Prediction values for screening tests of pulmonary function. *Am. Rev. Respir. Dis.*, 91, 252–61.

Gregg, I. and Nunn, A.J. (1973) Peak expiratory flow in normal subjects. *Br. Med. J.*, 3, 282–4.

Johannsen, Z.M. and Grasmus, L.D. (1968) Clinical spirometry in normal Bantu. *Am. Rev. Respir. Dis.*, 97, 585–92.

Juhl, B. (1970) Pulmonary function investigations on 1011 schoolchildren using Wright's Peak Flow Meter. *Scand. J. Clin. Lab. Invest.*, 25, 355–61.

Kallqvist, I., Taube, A., Olafsson, O. *et al.* (1970) Peak expiratory flow in healthy persons aged 45–65 years. *Scand. J. Respir. Dis.*, 51, 177–87.

Lam, K-K., Pang, S.C., Allen, W.G.L. *et al.* (1982) A survey of ventilatory capacity in Chinese subjects in Hong Kong. *Ann. Hum. Biol.*, 9, 459–72.

Leiner, G.C., Abramowitz, S., Small, M.J., Stinby, V.B. and Lewis, W.A. (1963) Expiratory peak flow rate. Standard values for normal subjects. Use as a clinical test of ventilatory function. *Am. Rev. Respir. Dis.*, **88**, 644–51.

Malik, S.K. and Banga, N. (1978) Peak expiratory flow rate in non-smoking rural males. *Ind. J. Chest Dis. All. Sci.*, **20**, 183–6.

Malik, M.A., Moss, E., Lee, W.R. *et al.* (1972) Prediction values for the ventilatory capacity in male W. Pakistani workers in the UK. *Thorax*, 27, 611–19.

Nairn, I.R., Bennett, A.J., Andrew, J.D. and Macarhew, P. (1961) A study of respiratory function in normal school children. The peak flow rate. *Arch. Dis. Child.*, **36**, 253–8.

Nunn, A.S. and Gregg, I. (1989) New regression equations for predicting peak expiratory flow in adults. *Br. Med. J.*, **298**, 1068–70.

Pelzer, A.M. and Thompson, M.L. (1964) Expiratory peak flow. *Br. Med. J.*, 2, 123.

Primhak, R.A., Biggins, S.D., Tsanakas, J.N. *et al.* (1984) Factors affecting the peak expiratory flow rate in children. *Br. J. Dis. Chest*, **78**, 26–35.

Ray, D., Rajaratnam, A., Richard, J. *et al.* (1993) Peak expiratory flow in rural residents of Tamil Nadu, India. *Thorax*, **48**, 163–6.

Appendix F Agents causing asthma in selected occupations

This is not a comprehensive list, but gives the main work areas and agents.

Occupation or occupational field	*Agent*
	Biological proteins
Laboratory animal workers, veterinarians	Dander and urine proteins
Food processing	Shellfish, egg proteins, pancreatic enzymes, papain, amylase
Dairy farmers	Storage mites
Poultry farmers	Poultry mites, droppings and feathers
Granary workers	Storage mites, aspergillus, indoor ragweed, grass pollen
Research workers	Locusts
Fish food manufacturing	Midges
Detergent manufacturing	*Bacillus subtilis* enzymes
Silk workers	Silk-worm moths and larvae
	Plant proteins
Bakers	Flour, amylase
Food processing	Coffee bean dust, meat tenderizer (papain), tea
Farmers	Soy bean dust
Shipping workers	Grain dust (moulds, insects, grain)
Laxative manufacturing	Ispaghula, psyllium
Sawmill workers, carpenters	Wood dust (Western Red Cedar, oak, mahogany, zebrawood, redwood, Lebanon cedar, African maple, Eastern White Cedar)
Electric soldering	Colophony (pine resin)
Cotton textile workers	Cotton dust
Nurses	Psyllium, latex
	Inorganic chemicals
Refinery workers	Platinum salts, vanadium
Plating	Nickel salts

Agents causing asthma in selected occupations (*Continued*)

Occupation or occupational field	
Diamond polishing	Cobalt salts
Manufacturing	Albuminium fluoride
Beauty shop	Persulphate
Welding	Stainless steel fumes, chromium salts
	Organic chemicals
Manufacturing	Antibiotics, piperazine, methyl dopa, salbutamol, cimetidine
Hospital workers	Disinfectants (sulphathiazole, chloramine, formaldehyde, glutaraldehyde)
Anaesthesiology	Enflurane
Poultry workers	Aprolium
Fur dyeing	Paraphenylene diamine
Rubber processing	Formaldehyde, ethylene diamine, phthalic anhydride
Plastics industry	Toluene diisocyanate, hexamethyl diisocyanate, dephenylmethyl isocyanate, phthalic anhydride, triethylene tetramines, trimellitic anhydride, hexamethyl tetramine
Automobile painting	Dimethyl ethanolamine diisocyanates
Foundry worker	Reaction product of furan binder

Appendix G Peak flow charts

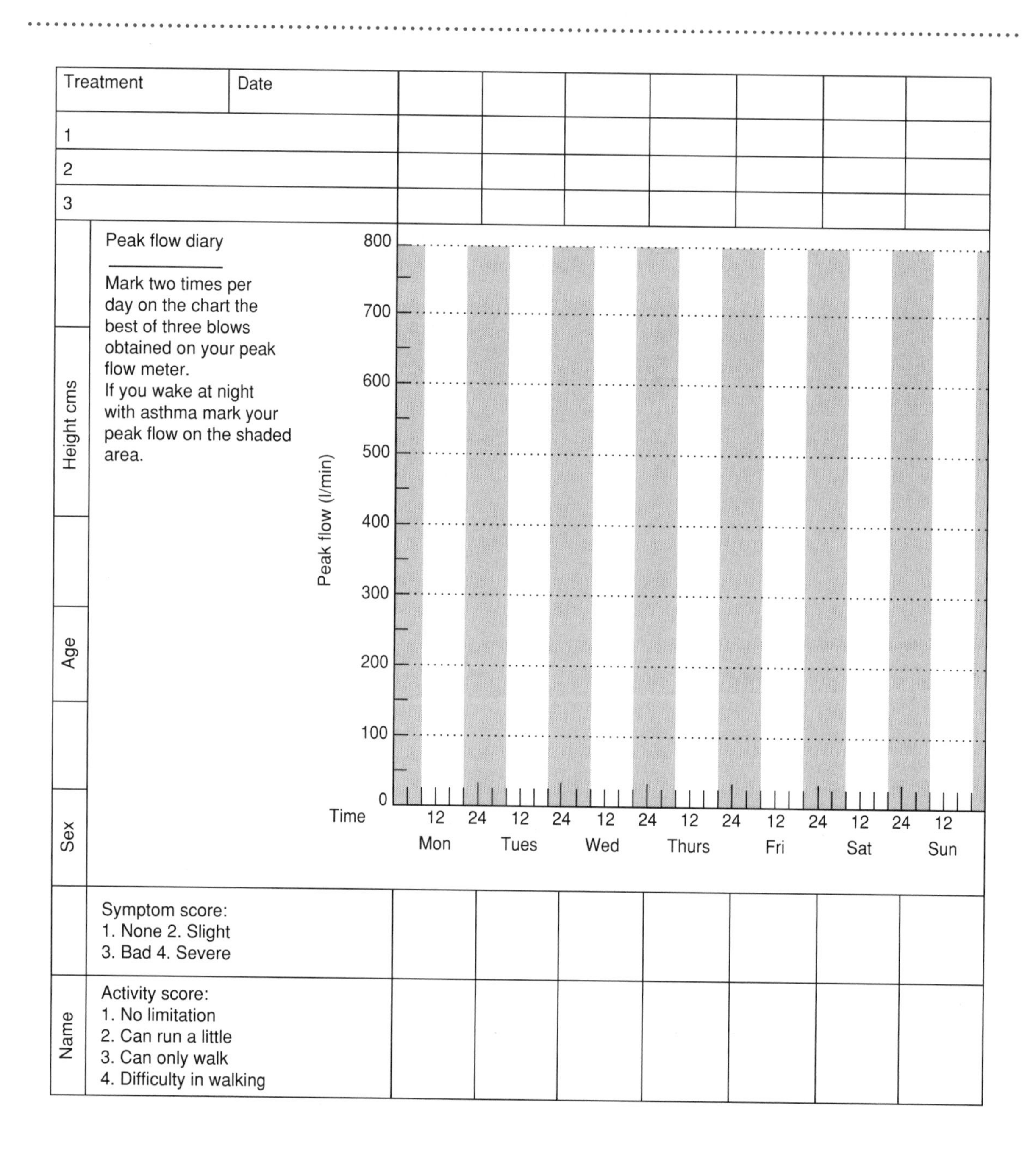

Treatment	Date							
1								
2								
3								

Height cms

Age

Sex

Name

Peak flow diary

Mark two times per day on the chart the best of three blows obtained on your peak flow meter.
If you wake at night with asthma mark your peak flow on the shaded area.

Peak flow (l/min)

800
700
600
500
400
300
200
100
0

Time

12 24 12 24 12 24 12 24 12 24 12 24 12

Mon Tues Wed Thurs Fri Sat Sun

Symptom score: 1. None 2. Slight 3. Bad 4. Severe							
Activity score: 1. No limitation 2. Can run a little 3. Can only walk 4. Difficulty in walking							

Treatment	Date							
1								
2								
3								

Height cms

Age

Sex

Name

Peak flow diary

Mark four times per day on the chart the best of three blows obtained on your peak flow meter.
If you wake at night with asthma mark your peak flow on the shaded area.

Peak flow (l/min): 800, 700, 600, 500, 400, 300, 200, 100, 0

Time: 12 24 12 24 12 24 12 24 12 24 12 24 12

Mon Tues Wed Thurs Fri Sat Sun

Symptom score: 1. None 2. Slight 3. Bad 4. Severe							
Activity score: 1. No limitation 2. Can run a little 3. Can only walk 4. Difficulty in walking							

Appendix H Table of percentage maximum peak flows for use in management plans

Peak flows	40%	50%	70%	80%	100%
200	80	100	140	160	200
210	84	105	147	168	210
220	88	110	154	176	220
230	92	115	161	184	230
240	96	120	168	192	240
250	100	125	175	200	250
260	104	130	182	208	260
270	108	135	189	216	270
280	112	140	196	224	280
290	116	145	203	232	290
300	120	150	210	240	300
310	124	155	217	248	310
320	128	160	224	256	320
330	132	165	231	264	330
340	136	170	238	272	340
350	140	175	245	280	350
360	144	180	252	288	360
370	148	185	259	296	370
380	152	190	266	304	380
390	156	195	273	312	390
400	160	200	280	320	400
410	164	205	287	328	410
420	168	210	294	336	420
430	172	215	301	344	430
440	176	220	308	352	440
450	180	225	315	360	450
460	184	230	322	368	460
470	188	235	329	376	470
480	192	240	336	384	480
490	196	245	343	392	490
500	200	250	350	400	500
510	204	255	357	408	510
520	208	260	364	416	520

Table of percentage maximum peak flows for use in management plans (*Continued*)

Peak flows	*40%*	*50%*	*70%*	*80%*	*100%*
530	212	265	371	424	530
540	216	270	378	432	540
550	220	275	385	440	550
560	224	280	392	448	560
570	228	285	399	456	570
580	232	290	406	464	580
590	236	295	413	472	590
600	240	300	420	480	600
610	244	305	427	488	610
620	248	310	434	496	620
630	252	315	441	504	630
640	256	320	448	512	640
650	260	325	455	520	650
660	264	330	462	528	660
670	268	335	469	536	670

Appendix I Useful addresses

Action Asthma
c/o Allen & Hanburys Ltd
Stockley Park West
Uxbridge
Middlesex
UB11 1BT

Tel. 0181 990 9888

British Lung Foundation (BLF)
New Garden House
78 Hatton Garden
London
EC1N 8JR

Tel. 0171 831 5831

British Thoracic Society (BTS)
6th Floor, North Wing
New Garden House
78 Hatton Garden
London
EC1N 8JR

Tel. 0171 831 8778

Clement Clarke International Ltd
Airmed House
Edinburgh Way
Harlow
Essex CM20 2ED

Tel. 01279 414969

Ferraris Medical Limited
Ferraris House
Aden Road
Enfield
Middlesex
EN3 7SE

Tel. 0181 805 9055

GPs in Asthma Group
Medical Marketing Interface
Bass Brewery
Tollbridge Road
Bath
Somerset
BA2 7DE

Tel. 01225 858880

Health & Safety Executive
Magdalen House
Stanley Precinct
Bootle
Merseyside
L20 3QZ

Tel. 0151 951 4000

Healthscan Products Inc.
908 Compton Avenue
Cedar Grove
New Jersey 07009–1292
USA

Tel. 1-800-962-1266 (in New Jersey: 201-857-3414)

Medix Ltd
Medix House
Catthorpe
Lutterworth
Leicestershire
LE17 6DB

Tel. 01788 860366

Micro Medical Ltd
P O Box 6
Rochester
Kent
ME1 1AZ

Tel. 01634 843383

National Asthma Campaign (NAC)
Providence House
Providence Place
London N1 ONT

Tel. 0171 226 2260

Royal College of Nursing
Respiratory Nurse Forum
RCN
20 Cavendish Square
London
WIM OAS

Tel. 0171 409 3333

Vitalograph Ltd
Maids Moreton House
Buckingham
MK18 1SW

Tel. 01280 822811

Index

Page numbers appearing in **bold** refer to figures and page numbers appearing in *italics* refer to tables